Electrothermal Atomization

for Atomic Absorption Spectrometry

Analytical Sciences Monographs

Electrothermal Atomization for Atomic Absorption Spectrometry

C. W. Fuller

Tioxide International Ltd
Cleveland, England

The Chemical Society
Burlington House
London W1V 0BN

ISBN: 0 85186 777 4
ISBN: 0583-8894

Typeset in IBM Press Roman by Preface Ltd, Salisbury, Wilts
and printed in Great Britain by offset lithography
by Billing & Sons Ltd, Guildford, London, and Worcester

Foreword

Since the introduction of atomic absorption spectrometry as an analytical technique, by Walsh, in 1953, the use of alternative atomization sources to the flame has been explored. At the present time the two most successful alternatives appear to be the electrothermal atomizer and the inductively-coupled plasma. Electrothermal atomizers have provided a considerable improvement in sensitivity for the majority of the elements determined by atomic absorption. This is brought about essentially by introducing the total quantity of sample to the atomizer, thereby overcoming the inefficiency of a nebulization system and avoiding the large dilution of the atomic population in the flame gases. Inductively-coupled plasmas have provided a higher temperature atomizer, thereby enabling the more efficient atomization of many elements. Their development, however, has occurred almost exclusively in the field of atomic emission spectrometry, where, with the higher temperatures available and the relatively low background, the possibilities of simultaneous multi-element analysis have increased considerably.

In this book an attempt has been made to provide the author's views on the historical development, commercial design features, theory, practical considerations, analytical parameters of the elements, and areas of application of the first of these two techniques, electrothermal atomization.

Contents

Foreword v

1 History 1

2 Electrothermal Atomizers 11

2.1 Commercial Designs 11
    (a) Graphite furnace atomizer 11
    (b) Graphite cup atomizer 14
    (c) Graphite rod atomizer 15
    (d) Tantalum filament atomizer 16
2.2 Pyrolytic Coatings on Graphite Components 17
2.3 Advantages and Disadvantages of Electrothermal
Atomization Compared with Flame Atomization 18
    (a) Increased sensitivity 18
    (b) *In situ* sample treatment 19
    (c) Small samples 20
2.4 Comparison of Detection Limits for the Various
Types of Electrothermal Atomizer 21

3 Theoretical Aspects of the Atomization Process 22

3.1 Thermodynamic 22
    (a) Evaporation of the metal oxide prior to
atomization 22
    (b) Thermal dissociation of the metal oxide 22
    (c) Reduction of the metal oxide 23
    (d) Carbide formation 26
    (e) Analytical applications of thermodynamic theories
of atomization 26
3.2 Kinetic 27
    (a) Atomization under increasing temperature 27
    (b) Atomization under isothermal conditions 31
    (c) Analytical applications of kinetic theories of
atomization 31

4 General Experimental Conditions   **41**

   4.1   Sample Handling   41
       (a) Solids   42
       (b) Liquids   43
       (c) Gases   47
       (d) Radioactive samples   47
   4.2   Separation Procedures   47
   4.3   Selection of Instrumental Parameters   49
       (a) Light sources   49
       (b) Atomizer parameters   50
       (c) Sheathing gas   52
   4.4   Background Correction   56
   4.5   Calibration   59
   4.6   Interferences   61
       (a) Physical interferences   61
       (b) Chemical interferences   62

5 Analytical Conditions for the Determination of the Elements by Atomic Absorption Spectrometry using Electrothermal Atomization   **65**

6 Applications   **84**

   6.1   Oil and Oil Products   84
   6.2   Metals   86
   6.3   Rocks, Minerals, and Soils   88
   6.4   Waters   92
   6.5   Plants   97
   6.6   Food and Drugs   97
   6.7   Biological Fluids   97
   6.8   Biological Tissues   103
   6.9   Air Particulates   105
   6.10 Refractory Oxides and Related Materials   107
   6.11 Other Analytical Applications   109
   6.12 Theoretical   110

References   **112**

Index   **125**

1 History

The development of electrothermal atomizers in the period 1959 to 1976 can be traced back to the work of A. S. King in 1905 and 1908. King was interested in the observation of emission spectra of elements without the complicating factors of electrical conduction, occurring in the atomic vapour produced by an arc or spark, and unknown chemical reactions, occurring in the atomic vapour produced by a flame. His aim was to obtain emission spectra produced, as nearly as possible, solely by the effect of heat.

King first used an arc-heated furnace,[1] shown schematically in Figure 1-1. The apparatus consisted of a carbon rod of 16 mm diameter (a) with a 5 mm diameter hole drilled through; this tube furnace was held inside a carbon block (b) but insulated from it by two tubes of asbestos (c) placed at each end. A hole was drilled out from the centre of the carbon block, at the base, to allow a carbon electrode (d) to enter. Copper clamps fitted to one end of the horizontal tube furnace and the vertical carbon electrode enabled the apparatus to be connected to a power supply. Samples were placed at the centre of the furnace and the electrode was raised until it formed an arc (up to 30 A at 220 V) with the furnace wall. Using this design, temperatures of 2200 °C were obtained. By extending one of the asbestos insulating tubes and fitting a quartz window (e) and then using an asbestos plug to fill the other end of the furnace, it was possible to operate in an almost enclosed environment.

After an initial design for a resistively heated carbon furnace,[1] King[2] went on to use the carbon furnace shown schematically in Figure 1-2. This apparatus consisted of a carbon tube (a) 150–200 mm in length, of 16 mm external diameter and 4 mm internal diameter. The wall thickness was reduced to 0.5 mm over the length of the tube, apart from 15 mm at each end. This carbon tube was fixed along the centre axis of a brass cylinder (b), 400 mm in length and 100 mm in diameter, by two supports (c), one fitted directly to the cylinder and the other insulated from it. One end of the brass cylinder was permanently closed and fitted with a quartz window (d) while the other end was fitted with a removable cap also fitted with a quartz window (d). Electrical power, 5 kW, was supplied to the furnace through the two copper leads (e). The furnace and chamber could be evacuated and filled with hydrogen through (f).

King continued to use his electrically heated furnaces to investigate the emission spectra of many elements. No use was made of the principle of electrothermal atomization for quantitative analytical measurements, however, until 1956, when two groups of workers in the U.S.S.R. used an electrothermal vaporization technique to separate and preconcentrate trace elements onto a graphite electrode prior to emission spectrochemical

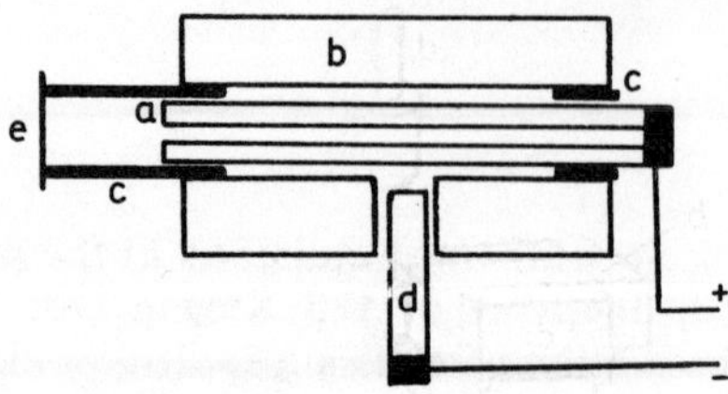

Figure 1-1   Arc atomizer designed by King.

analysis. The apparatus used by Mandelshtam, Semenov, and Turovtseva[3] is shown schematically in Figure 1-3. The sample, in the range 1–1000 mg, was placed in the graphite cup (a), which was held between two graphite brushes (b). The graphite brush holders (c) also carried cooling water and were inputs for the electrical power. One of these graphite brushes could be moved to allow the sample cups to be positioned, adjusted, and removed, as necessary. A uniform pressure was exerted on the sample cup by means of a spring and an adjusting screw. The brush holders were connected to the low-voltage winding of a 5 kW step-down transformer, which, using a variac in the transformer primary, enabled the cup to be heated to temperatures in the region of 1800 °C. Impurities, in samples of low volatility, were evaporated from the cup and collected on a graphite receiver electrode fitted into the holder (d). The evaporated material was then excited in an arc or a spark to obtain the emission spectra. The technique used by Zaidel, Kaliteevskii, Lipis, Chaika, and Belyaev[4] was very similar, except that their evaporator was operated in an enclosed system, under vacuum. It was several years before Belyaev and co-workers used a version of this apparatus for atomic fluorescence and atomic absorption measurements.[52]

Three years later, in 1959, L'vov began to publish his classic work on the application of electrothermal furnace atomizers for quantitative atomic absorption analyses.[5] The graphite furnace used in his early work,

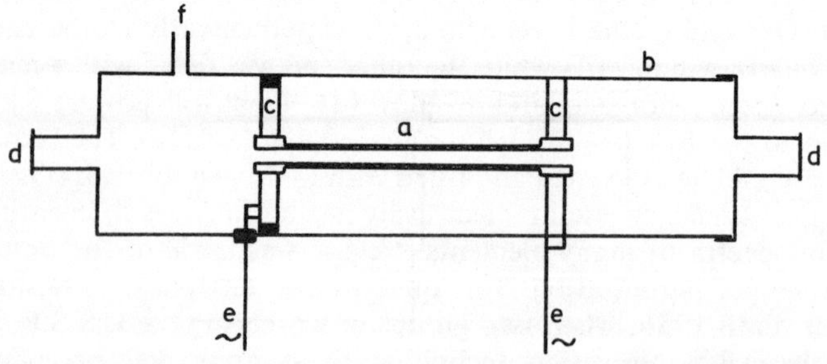

Figure 1-2   Electrothermal atomizer designed by King.

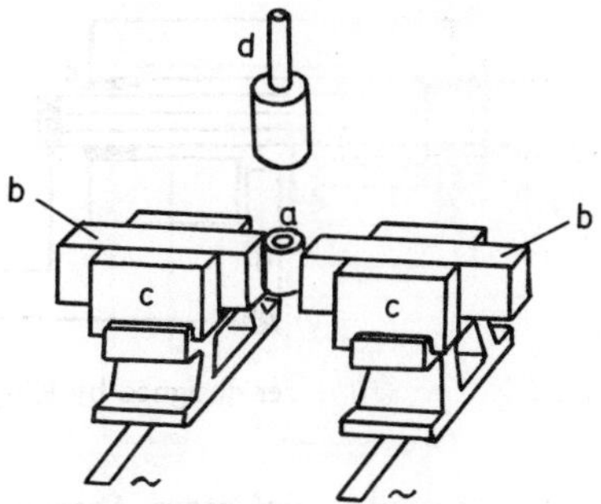

Figure 1-3 Electrothermal vaporizer designed by Mandelshtam, Semenov, and Turovtseva.

similar in principle to the arc atomizer of King, is shown schematically in Figure 1-4. The apparatus consisted of a graphite furnace (a), 100 mm in length, of 10 mm external diameter and 3 mm internal diameter, lined with tantalum foil (b), to decrease the loss of atomic vapour by diffusion through the porous carbon, a carbon electrode (c) for sample introduction, and an auxiliary electrode (d) for arcing (d.c.) to the sample electrode. This furnace was contained under an aluminium cover containing two quartz windows for the atomic absorption measurements, and one window for observation. Analyses were carried out by placing samples, 0.1 mg, on the tip of the sample electrode; four sample electrodes could be accommodated in the apparatus, which was then evacuated and filled with argon. The carbon furnace was heated to the desired temperature, using up to 10 kW power, before a sample electrode was moved into the opening in the furnace wall. As the sample entered the furance a d.c. arc was automatically switched on for 3—4 seconds between the tip of the sample electrode and the auxiliary electrode, causing the sample electrode tip to heat and sample vaporization and atomization to occur.

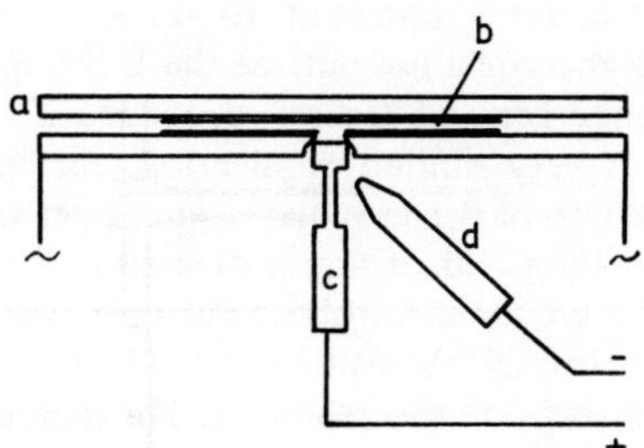

Figure 1-4 Arc atomizer designed by L'vov.

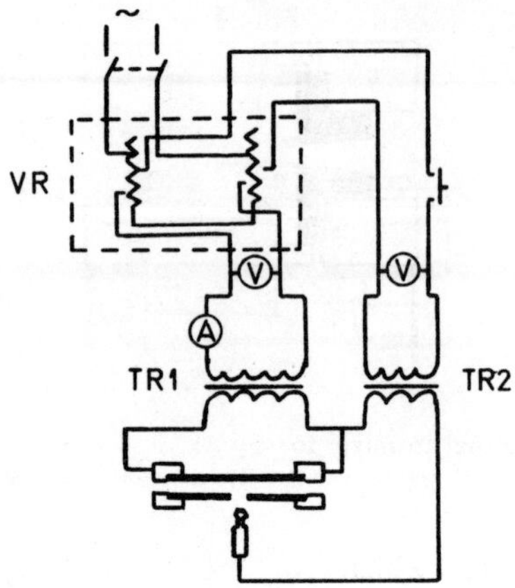

Figure 1-5   Electrothermal atomizer designed by L'vov.

Heating the sample *via* the d.c. arc, however, proved to be inefficient, and a later version of the graphite furnace,[6] in 1967, utilized resistive heating for the sample electrode as well as for the furnace. The circuitry for this design is shown in Figure 1-5; where VR represents the voltage regulator to the two step-down transformers TR1 (4 kW) and TR2 (1 kW) which were used for heating the furnace and sample electrode, respectively. In this method the electrode tip was heated by an alternating current passing through the furnace and the electrode; since the highest resistance occurred at the contact and the electrode neck, it was mainly the electrode head that was heated. The dimensions of this furnace were smaller than in the earlier version; the graphite cylinders measured 30—50 mm in length with an internal diameter of 2.5—5.0 mm and an external diameter of 6.0 mm. The graphite cylinders were no longer lined with tantalum or tungsten foil to prevent vapour diffusion; instead the more practical solution of using a pyrolytic graphite coating was employed. L'vov's apparatus, while giving the best absolute sensitivities for atomic absorption analyses obtained to date,[7] has not been readily accepted for routine analytical use outside the U.S.S.R.

In 1967, Brandenberger and Bader[8] produced a copper wire electrothermal atomizer, of very limited application, for the determination of mercury. A 0.4 m length of 0.1 mm diameter copper wire was formed into a coil 10 to 15 mm long and of 4 mm diameter. This coil was used as a cathode to deposit mercury electrolytically onto the wire before it was positioned in a flow-through absorption cell, which was held in the optical path of an atomic absorption spectrometer. The mercury was atomized by heating the coil electrically.

The year 1967 saw the biggest step forward in the development of an

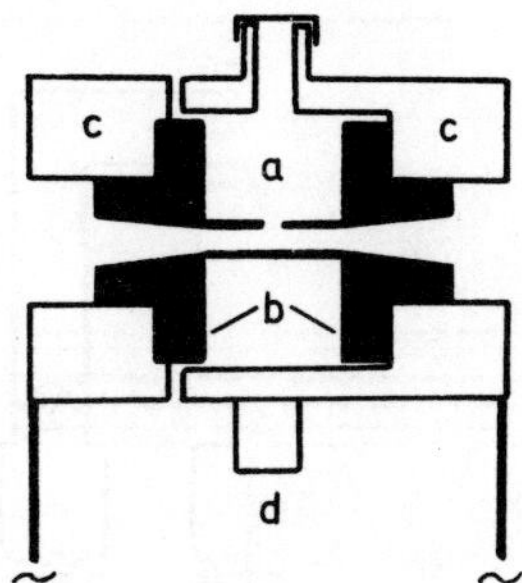

Figure 1-6   Electrothermal atomizer for atomic absorption measurements, designed by Massmann.

electrothermal atomizer which could be used in a routine analytical laboratory with commercially available atomic absorption spectrometers. Massmann produced his much simplified and compact versions of the resistively heated King furnace to enable both atomic absorption and atomic fluorescence measurements to be made.[9,10] The graphite furnace for atomic absorption analyses, Figure 1-6, consisted of a graphite tube (a) 55 mm in length, of 6.5 mm internal diameter and 8.0 mm external diameter; a hole, diameter 2 mm, was drilled through the wall at the centre of the tube for the introduction of liquid samples, 5–200 μl. The graphite tube, operated in an atmosphere of argon, was supported between two water-cooled steel cones (b) which were electrically insulated from each other; the graphite tube was held in a metal casing (c) which was supported on the mounting (d). The graphite furnace for atomic fluorescence analyses, Figure 1-7, consisted of a graphite cup (a), 40 mm high, of 6.5 mm internal diameter and 8.0 mm external diameter, having a sample capacity of 5 to 50 μl, with slits cut into both sides for fluorescence

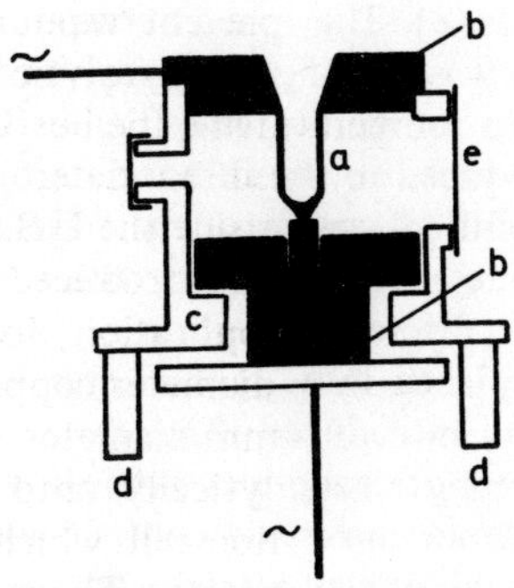

Figure 1-7   Electrothermal atomizer for atomic fluorescence measurements, designed by Massmann.

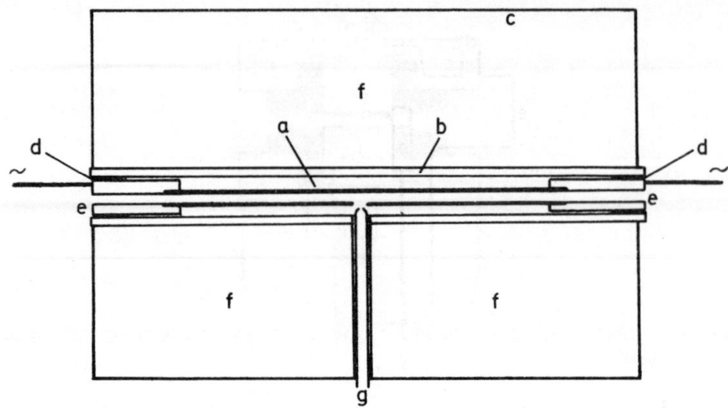

Figure 1-8   Electrothermal atomizer designed by Woodriff and Ramelow.

measurements. The other components corresponded to those described for the atomic absorption furnace with the exception that the atomic fluorescence signal was measured through the quartz window (e). Irradiation of the atomic vapour occurred by focusing light from a hollow-cathode lamp vertically downwards into the graphite cup.

Also around this time Woodriff introduced his furnace design for atomic absorption analysis;[11,12] a design similar in some respects to that of L'vov. Woodriff's initial apparatus[11] is shown in Figure 1-8; this early design was modified and improved several times on later occasions. The furnace consisted of a graphite tube (a), 250 mm long, of 6 mm internal diameter and 8 mm external diameter, supported in a larger graphite tube (b), of external diameter 25 mm, which was held in a stainless-steel housing (c). The two graphite tubes were separated by a layer of insulation (d) at each end. Power was fed to the graphite furnace through the two graphite supports (e). The space between the outer graphite tube and the metal housing was filled with graphite-felt insulation (f). The furnace was preheated to temperatures of 3000 °C before liquid samples were nebulized and introduced to the furnace, through the side-arm (g), by a stream of argon. A modified version of the furnace illustrated, where the absorption tube was divided into two parts with a thick supporting tube in the centre, was used for discrete sampling of liquids and solids. In this apparatus samples were placed in a graphite cup, of 50 μl capacity, on the end of a graphite rod and introduced through the side-arm. The high heat capacity of the centre support tube was sufficient to vaporize the sample completely when the sample cup made contact with it. Argon was passed around the graphite rod as it was introduced, so as to prevent air from entering the graphite tube. The large size of the Woodriff furnace has prevented its use in commercial atomic absorption spectrometers and it

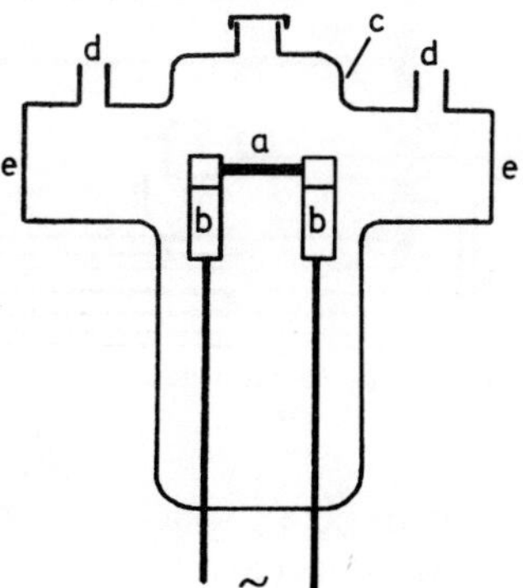

Figure 1-9   Carbon rod atomizer designed by West and Williams.

has therefore not received the same wide acceptance as the smaller Massmann furnace.

A further important development occurred in 1969 when West and Williams[13] published their work on a carbon filament atomizer for atomic absorption and atomic fluorescence measurements. The atomic absorption apparatus, shown schematically in Figure 1-9, was very simple in design, consisting of a carbon filament (a), 40 mm long, between 1 and 2 mm diameter and of sample capacity 5 μl, supported on two stainless-steel electrodes (b); the atomizer was enclosed in a Pyrex glass chamber (c) with a stream of argon (d) flowing through it. Two silica windows (e) in the glass chamber were used to enable light from a hollow-cathode lamp to pass through the chamber and just above the surface of the carbon filament. The atomic fluorescence system was identical except that the two side-arms of the glass chamber were positioned at 90° to each other to allow irradiation of the atomic vapour at right angles to the optical path. This atomizer design was simplified further in 1970 by removing the surrounding glass chamber and using a vertical flow of inert gas around the carbon filament.[14] The power requirements of the filament atomizer were considerably lower than the earlier tube atomizers; 0.5 kW compared with 5 kW.

The development of metal filament atomizers also continued in 1969 with the use of an electrically heated platinum or tungsten wire-loop atomizer for atomic fluorescence measurements.[15] The real development in the field of metal filament atomizers, however, came a year later with the electrically heated tantalum boat developed by Donega and Burgess,[16] shown in Figure 1-10. The chamber (a) was constructed from a quartz tube of 50 mm internal diameter, with an optical quartz window (b) sealed on one end. The other end of the quartz tube was sealed with a brass plate (c) which contained a second quartz window (b), a gas/vacuum

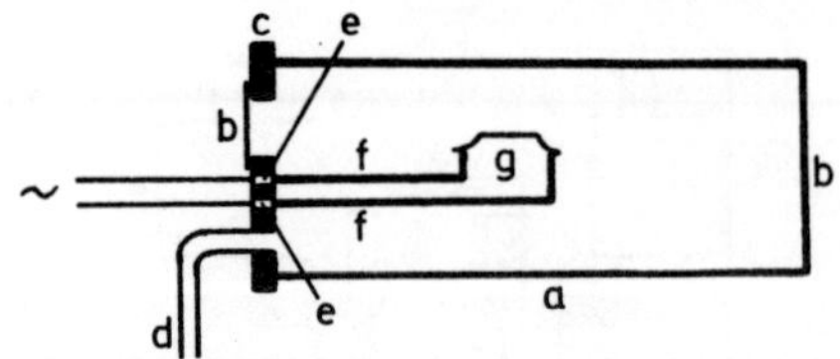

Figure 1-10   Tantalum boat atomizer designed by Donega and Burgess.

port (d), and two insulator rings (e), which held two 6.35 mm diameter copper rods (f). The sample boats (g), 50 mm long, 6 mm wide, and of sample capacity 100 μl, were held on the copper rods by two small clamps. The sample boats were manufactured either from tantalum or tungsten foil or from graphite sheet. The chamber was evacuated and refilled with an inert gas, if required. Atomizer temperatures up to 2200 °C could be obtained with a power requirement of only 0.5 kW. A disadvantage of this, and other enclosed systems, was that material vaporized from the atomizer rapidly condensed within the chamber and on the quartz windows.

The equipment described above represents the position in the development of basic design principles for electrothermal atomizers at the present time. There have been many suggested variations of these designs, however, of which the following deserve mention. Montaser, Goode, and Crouch,[17] in 1974, proposed the use of a flexible graphite braid filament, 30 mm long and of between 1.5 and 2.0 mm diameter, as a replacement for the rigid graphite rods used with other filament designs. Also in 1974, Katskov, Kruglikova, L'vov, and Polzik[18,18a] published a modified design of the Massman type furnace, which enabled sample weights of up to 100 mg to be analysed. Their furnace, shown schematically in Figure 1-11, consisted of two coaxially clamped graphite cylinders, 15 mm long and of internal diameters 3 mm (a) and 7 mm (b). The sample was introduced between the two cylinders and the atomic vapours produced on heating to temperatures of 2600 °C diffused through the walls of the inner graphite cylinder into the optical path of the spectrometer.

Robinson and Wolcott,[19] in 1975, extended their work with r.f.-heated

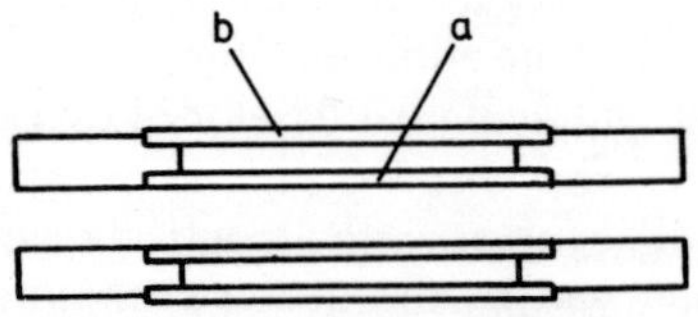

Figure 1-11   Double furnace designed by Katskov, Kruglikova, L'vov, and Polzik.

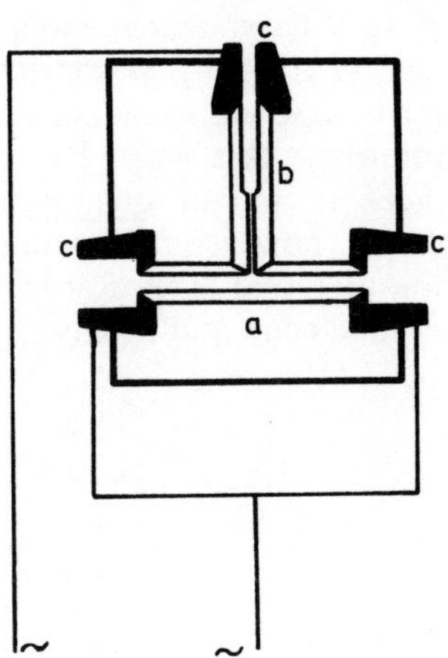

Figure 1-12   T-shaped electrothermal atomizer designed by Robinson and Wolcott.

atomizers to produce a compact T-shaped electrothermal carbon furnace, similar in design to a combined Woodriff and L'vov furnace. The design, shown schematically in Figure 1-12, consisted of a horizontal graphite tube (a), 75 mm long, of 17.5 mm external diameter and 4.8 mm internal diameter, and a vertical graphite tube (b), 64 mm long, of 17.5 mm external diameter and 6.2 mm and 1.6 mm internal diameters at the top and bottom sections, respectively. The two graphite tubes were fitted together in the T-shape and supported by the three graphite end cones (c); both tubes were heated resistively by a 4.5 kW power supply. The furnace was preheated to temperatures up to 2600 °C before liquid or solid samples were introduced into the vertical graphite tube. The samples were vapourized and atomized and then swept into the horizontal graphite tube, by a stream of inert gas, where atomic absorption measurements were made.

Electrothermal atomizers have reached a point in their history where further refinements in design will undoubtedly continue but where the next major advance in design principle must come from a reconsideration of the atomization processes. Electrothermal atomizers are efficient in the process of sample vaporization but relatively inefficient in the process of atomization. To obtain a significant improvement in sensitivity it is necessary to overcome the problem, inherent in all electrothermal atomizers, of sample vaporization occurring without complete atomization.

It is interesting to note that, while the original work of King was solely concerned with the measurement of emission spectra, the analytical applications of electrothermal atomizers have utilized the technique of atomic absorption, and to a lesser extent atomic fluorescence, spectrometry. Some recent work, in 1974, by Massmann and Gucer[20] served to remind analytical chemists that emission measurements could be made with electrothermal atomizers. This work has since been extended by other groups of workers to obtain quantitative analytical measurements[21-22a] It is unlikely, however, that atomic emission will replace atomic absorption as the preferred analytical technique unless advantage can be taken of the simultaneous multi-element analytical capability of emission spectrometry.

2 Electrothermal Atomizers

In the previous chapter the historical development of electrothermal atomizers was outlined; in this chapter commercial versions of the carbon furnace, cup, and rod, and the tantalum filament, are used to illustrate atomizer design features.

The heating of all commercial atomizers is controlled by versatile power units which enable the selection of drying, pyrolysis, and atomization times and temperatures. The criteria involved in the selection of these parameters are outlined in Section 4.3(b). Most atomizer control units also have built-in safety devices which shut off the atomizer power supply should the inert-gas supply fail or the temperature of the atomizer housing become too high. Further details of these power supplies are not discussed here unless they are particularly relevant.

A complete list of currently available commercial atomizers, together with many of their salient features, is given each year in 'Annual Reports on Analytical Atomic Spectroscopy' (published by the Chemical Society), so no attempt has been made to list them here.

2.1 Commercial Designs
(a) Graphite Furnace Atomizer
The Perkin-Elmer Heated Graphite Atomizer (HGA) series was introduced in 1970 with a design based mainly on the work of Massmann.[10] The HGA 70, shown schematically in Figure 2-1, consists of a graphite cylinder (A), 51 mm long by 8.6 mm internal diameter, supported at each end by graphite cones (B), which are held in the water-cooled metal housing (C) of the atomizer unit. The graphite cones act as optical apertures by limiting the amount of light, from the heated tube, reaching the photomultiplier. Electrical power (10 V and 500 A maximum) is fed to the atomizer by the cables (D), enabling the graphite tube to be heated to temperatures of 2500 °C within 5 seconds. The graphite tube is protected from atmospheric oxidation by a stream of argon or nitrogen flowing round and through the tube. The inert-gas flow enters through the two holes tangentially opposing each other at either end of the graphite tube such that the entry of air to the atomizer is considerably reduced. The outside temperature of the atomizer is maintained below 60 °C by the flow of water through the metal housing (C). The whole atomizer is positioned in the atomic absorption spectrometer in place of the normal burner assembly, so that the light beam from the hollow-cathode lamp passes through the centre of the graphite tube.

Liquid samples (100 μl maximum) are introduced into the furnace through the centre hole in the graphite tube while solid samples are introduced through the open ends of the graphite tube.

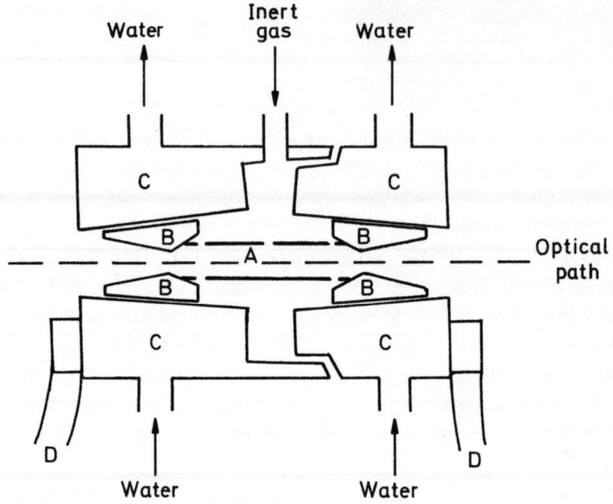

Figure 2-1   Schematic diagram of the Perkin-Elmer HGA 70.

Modifications to this early design have been made by several workers which have improved both the performance and the ease of operation of the equipment. The main problem with this early equipment was that the pyrolysis stages available with the control unit were limited to five preset temperatures (230, 330, 490, 750, and 1100 °C), and these temperatures were frequently inadequate for the analysis of some materials. Segar and Gonzalez[23] modified the circuitry of the control unit and introduced a variac, so that a completely variable pyrolysis temperature could be selected. Kahn and Slavin[24] described a modification to the inert-gas flow system, using a three-way solenoid valve, which allowed the flow of gas to be momentarily stopped during the atomization stage of a determination. This produced an enhancement in sensitivity for the determination of some elements by up to five times. Sperling[24a] has further modified this facility by redirecting the gas to the ends of the graphite tube during the normal gas-stop operation.

The design of the graphite tube was found to have limitations when analysing certain samples (particularly organic solvents) due to spreading of liquids along the length of the tube. A reduction in sensitivity and precision was produced, together with inefficient pyrolysis of the sample. A modified tube design was suggested[25] which incorporated a shallow groove (1 mm deep and 10 mm wide) at the centre of the inside surface of the tube, to retain the sample and prevent spreading, and a greater wall thickness at the centre, to produce a more even temperature along the tube. Although the maximum attainable temperature was reduced, the precision of the analytical determinations was significantly enhanced.

Issaq and Zielinski[26] have also suggested a modified graphite tube design which is claimed to reduce the problem of sample condensation at the cool ends of the graphite tube, which leads to severe light-scattering effects. In their design they cut two windows (7 mm x 13 mm) at both ends in the wall of a standard graphite tube. This modification (i) decreased the surface area of the graphite tube at the ends where sample condensation could occur, (ii) provided a more efficient flushing system, and (iii) produced a high temperature along a greater length of the graphite tube. While this tube design clearly has several advantageous features, it would be more difficult and more expensive to manufacture, and would certainly be less robust than the standard tube design.

Sperling improved the precision obtainable with some complex matrices by fitting a quartz window to one end of the graphite furnace.[26a] It was claimed that this improvement was caused by the modified gas-flow pattern produced.

Several authors have used a tantalum lining in graphite furnace atomizers to achieve increased sensitivity for some elements. These liners are not particularly successful, however, for routine analytical determinations. Coating the inside of a graphite tube with carbide-forming elements[27-28a] (*e.g.* B, Ba, La, Mo, Nb, Si, Ta, V, Zr) has been shown to enhance the analytical response significantly for some elements, and would seem to be a more practical method. This coating procedure is carried out by pyrolysing aqueous or organic solutions of the desired element or mixtures of elements introduced to the graphite furnace. Signal enhancement was found to be more pronounced for those elements which might be expected to form stable carbides.

When solutions containing, for example, sulphuric, perchloric, and hydrofluoric acids are analysed, corrosive acid fumes pour out from the ends of the graphite furnace during the pyrolysis stage, and can eventually cause quite severe corrosion of the surrounding metal components and the quartz windows of the spectrometer. Fuller[29] described a simple fume-extraction system to overcome this problem. Plastic cylinders were fitted to each end of the furnace and these were connected to a small vacuum pump *via* plastic tubing coming from the side-walls of the plastic cylinders. The pumping rate was controlled, and kept at the minimum level sufficient to extract the acid fumes efficiently.

The design of the HGA system has been modified several times since its introduction, incorporating several of the features described above. With production of the HGA 74, however, the only change in the basic concept of the original HGA system has occurred. This atomizer is smaller but the important feature is the reduction in size of the graphite tube itself. Analytical sensitivity is independent of tube length (provided the average residence time of atoms in the furnace is greater than or equal to the average atomization time) but inversely proportional to the cross-sectional area of the tube. Tube size in the HGA 74 was reduced to 28 mm length and 6.5 mm diameter, and the internal ends of the tube were grooved to retain sample solutions so that they could be used equally well with aqueous or organic solvents. The reduced size of the graphite tube enabled the high power requirements of the HGA 70 to be reduced considerably (10 V and 300 A maximum) and the highest temperature attainable to be increased to 2700 °C. The inert-gas flow system was changed, with the internal flow of gas through the graphite tube passing from the two cold ends to the hot central region; this modification has required that quartz windows be fitted to either end of the furnace to give a semi-enclosed configuration. This new flow pattern greatly reduced the problem of vapour condensation at the cool ends of the graphite tube that was so pronounced with the HGA 70 furnace design.

(b) Graphite Cup Atomizer.

The Varian Techtron Carbon Rod Atomizer (CRA) series has seen several design changes since its introduction in 1970. The design has changed from a graphite rod of the West and Williams style[13] to a graphite rod with a hole drilled through the centre to contain the sample (the mini-Massmann furnace), and then to a combined tube and cup atomizer; the CRA 63 or more recently the CRA 90. The cup atomizer is described here, although the only difference from the tube version is the replacement of the cup by a small horizontally placed tube furnace (9 mm long by 3 mm internal diameter). The CRA 63, shown schematically in Figure 2-2, closely resembles the atomizer design of Mandelshtam *et al.*[3] described in Chapter 1. It consists of a graphite cup (A) 9 mm high by 3 mm internal diameter supported between two graphite electrodes (B), which are held in the water-cooled metal mounting (C). Electrical power (10 V and 300 A maximum) is fed to the atomizer through the cables (D). The CRA 63 is operated in a stream of inert gas from the gas box (E). The cup is positioned in the optical path of the spectrometer so that light from the hollow-cathode lamp passes through the two small apertures in the top of the cup and through the external aperture (F). The size of the light aperture (F) is selected such that the minimum quantity of light, from the heated graphite cup, reaches the photomultiplier. A simple collimating lens system has been shown to improve the performance of spectrometers designed primarily for flame atomization.[29a]

Liquid (20 μl maximum) and solid samples are introduced directly through the open top of the graphite cup; the cup can then be heated rapidly (within 2 seconds) to temperatures of up to 3000 °C.

Due to repeated heating of the carbon cup, deterioration of the contacts on the graphite support electrodes occurs. Poor electrical contact results, giving rise to changes in atomization temperature and therefore to changes in reproducibility of analytical determinations. Johnson and Skogerboe[30] have described a simple rod device, machined from carborundum, with a diameter exactly equal to the external diameter of the

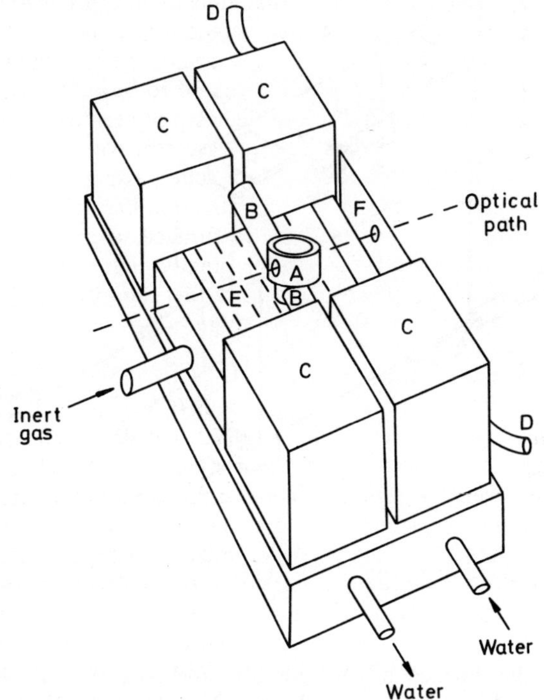

Figure 2-2   Schematic diagram of the Varian Techtron CRA 63.

cup, which can be used to regenerate the contacts on the support electrodes. This is performed by removing the carbon cup from between the support electrodes and inserting the carborundum rod in its place: the contact surfaces can then be re-ground by gently rotating the rod. Re-grinding of the electrode surfaces, after every 100–200 determinations, extends the useful life of the electrodes by a factor of five times.

(c) Graphite Rod Atomizer.
The rod represents the simplest design of the four types of atomizer described here, and is based on a simplified version of the carbon rod of West and Williams.[13] The Shandon Southern A3470 Flameless Atomizer, shown schematically in Figure 2-3, was introduced in 1971. It consists of a specially shaped carbon rod (A), 73 mm long by 3.1 mm diameter, clamped between two metal support pillars (B), which are air-cooled. Power (12 V and 100 A maximum) is fed to the atomizer through the cables (C). The atomizer is positioned in the optical path of the atomic absorption spectrometer so that the light from the hollow-cathode lamp

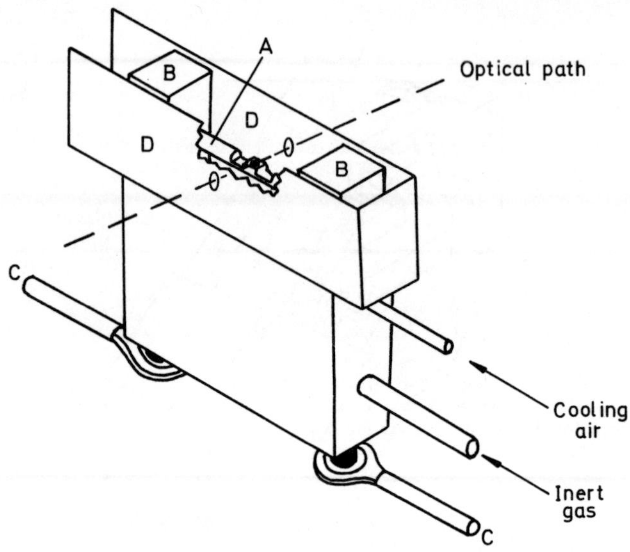

Figure 2-3   Schematic diagram of the Shandon Southern A3470.

passes over the surface of the graphite rod and through the external
apertures (D). As with the cup design, the optical baffle system (D) is
used to restrict the amount of light, from the heated graphite rod, reaching
the photomultiplier. These apertures also serve to restrict the area of
measurement to the position of maximum atom population, just above the
surface of the rod, thereby considerably reducing the interference
problems caused by vapour condensation in the cool region immediately
above the graphite rod.

Liquid samples (10 μl maximum) are pipetted into the small cup-shaped
indentation at the centre of the graphite rod. The graphite rod can then be
heated, in a stream of inert gas, to temperatures in excess of 3000 °C
within 2 seconds.

(d) Tantalum Filament Atomizer.

The Instrumentation Laboratory Flameless Sampler 355 was produced in
1971 with the design based on a modified version of the tantalum filament
atomizer of Donega and Burgess.[16] The IL 355 is shown schematically in
Figure 2-4. The atomizer consists of a bridge-shaped tantalum strip (A)
(14 mm wide and 58 mm long) supported by two metal clamps (B), fitted
inside a sealable housing (C) which can be purged with an inert gas. Power
(4 V and 100 A maximum) is supplied to the metal filament through the
cables (D). The atomizer is fitted into the optical path of the atomic

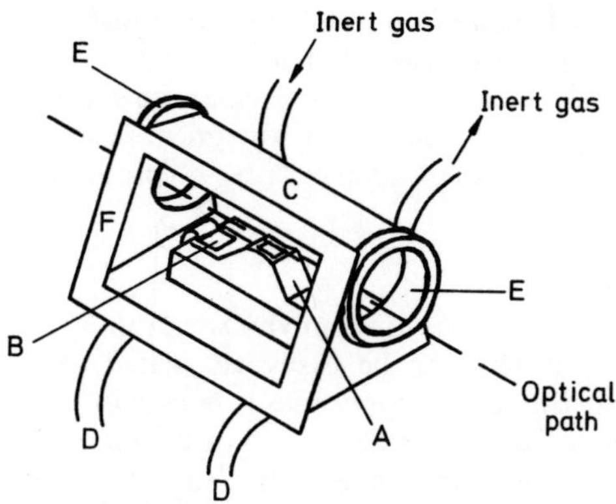

Figure 2-4   Schematic diagram of the Instrumentation Laboratory IL 355.

absorption spectrometer so that light from the hollow-cathode lamp passes through the two quartz windows (E) and just above the surface of the tantalum filament. Liquid samples (100 μl maximum) are pipetted on to the indentation at the centre of the filament by opening the access door (F) in the front of the atomizer housing. The tantalum filament atomizer can be heated to temperatures of 2400 °C (or 2800 °C, using a modified filament[31]) within 2 seconds.

2.2 Pyrolytic Coatings on Graphite Components

Some recent designs of commercial graphite atomizers incorporate components coated with pyrolytic graphite. However, these coatings can be removed with use, causing a reduction in performance characteristics; additionally, many atomizer designs have not used pyrolytic coatings. Siemer, Woodriff, and Watne[32] have described an extremely simple system for coating carbon components with pyrolytic graphite.

An insulated 17 mm Vycor tube, wound with about fifty turns of 18 gauge nichrome wire, was used as the sample chamber. Nitrogen was passed through the tube and the power to the heating wire (from a variable auto-transformer) was turned on, giving a temperature of 1000 °C. The object to be coated was introduced into the tube and the nitrogen flow replaced by 30–200 min^{-1} of methane, the flow rate being determined by the object. Depending on the size and shape of the component, the rate of deposition of pyrolytic graphite was from 0.01 to 0.1 mm per hour.

An additional and useful application of this technique is for joining graphite components. The components need to be placed in intimate contact within the tube and treated for one hour; they are then securely welded together.

2.3 Advantages and Disadvantages of Electrothermal Atomization Compared with Flame Atomization

There are three advantages generally claimed for electrothermal atomization; (*a*) greater sensitivity, (*b*) sample treatment *in situ*, and (*c*) capability to analyse small quantities of sample. These points are examined in greater detail.

(a) Increased Sensitivity

The theoretical improvement in sensitivity which should be attainable with electrothermal atomizers can be easily calculated. Assuming, for flame atomization, that the sample spray rate is 6 ml min^{-1}, that 10 per cent of the sprayed sample reaches the flame, and that the residence time of atoms in the analytical zone of the flame is 10^{-4} s,[7] then the volume of sample present in the analytical zone at any time is 10^{-6} ml. Assuming, for electrothermal atomization, that the production of atoms occurs as efficiently as in a flame and that the total sample is atomized before significant loss of the atomic population occurs from the atomizer, then for a 100 μl sample volume the improvement in sensitivity should be $100 \times 10^{-3}/10^{-6}$, or 100 000 times. If the same instrumentation is used for flame and electrothermal atomization then the noise levels in the two determinations will be very similar, and therefore the detection limits for the two techniques can be compared (Table 2.1). These results show that although a considerable improvement in detection limits is obtained, the levels of improvement are well below the theoretical value calculated above. These results are typical of the values obtained for other types of atomizer. Two conclusions can be reached from the results; either the atomization process is considerably less efficient in electrothermal atomizers than in flames, or the production of atoms in the atomizer is slow compared to their rate of removal. Both these conclusions lead to the assumption that there is considerable scope for improvement in the design of electrothermal atomizers.

Although no work has been reported on the design efficiency of furnace atomizers, this assumption is supported by the work of Cresser and Mullins,[33] who have studied the design of carbon- and tungsten-rod atomizers. These authors used the premise that the rate of increase of temperature of the atomizer is an important factor in determining the efficiency of the atomization process. Carrying out a theoretical study of the variation in atomizer heating rate as a function of various atomizer parameters, they reached the following conclusions for the optimum design features of carbon- and tungsten-rod atomizers: (*a*) doubling the applied voltage quadruples the heating rate; (*b*) for a fixed applied voltage and rod length, higher rod diameters are advantageous, owing to the reduction in heat losses by radiation; (*c*) for a fixed applied voltage and rod diameter, short rod lengths are considerably better, owing to the reduced power requirements to reach a fixed temperature; (*d*) for a fixed

Table 2.1 A comparison of detection limits for flame and electrothermal atomization (data obtained from Perkin-Elmer literature)

| Element | Detection limit [a]/μg l^{-1} | | Ratio of detection limits Flame : HGA 74 |
	HGA 74[b]	Flame	
Ag	0.001	2	2000
Al	0.02	30	1500
Au	0.1	20	200
Ba	0.5	20	40
Bi	0.2	40	200
Cd	0.001	1	1000
Co	0.05	10	200
Cr	0.1	3	30
Cu	0.02	2	100
Fe	0.03	10	330
Mn	0.002	2	1000
Mo	0.03	30	1000
Ni	0.1	10	100
Pb	0.02	20	1000
Pt	2	100	50
Si	0.5	80	160
Sr	0.05	10	200
Tl	0.1	30	300
V	1	60	60
Zn[c]	0.0005	2	4000

(a) Measurements taken with a PE 503 atomic absorption spectrometer.
(b) 100 μl sample volume.
(c) Extrapolated detection limit.

voltage and rod dimensions, tungsten is the preferred material rather than graphite; and (e) owing to the short atomization periods for rods, high-frequency modulation of light sources and detectors should be used, or alternatively a d.c light source and detector system should be installed. Cresser and Mullins suggest that a highly efficient atomizer could be designed incorporating d.c. electronics, and heating the atomizer by using a fixed quantity of charge from a capacitor.

(b) In situ *Sample Treatment*
Selective volatilization and matrix modification within the atomizer, together with the ability to analyse (directly) viscous liquids, are important advantages, and the techniques have been used frequently for organic matrices and occasionally for inorganic matrices. *In situ* sample treatment, however, could be regarded as a necessity of this technique, rather than an advantage, owing to the inability of electrothermal atomizers to accommodate high matrix concentrations during atomization.

(c) Small Samples
The facility to analyse small quantities of sample is an advantage only if limited sample quantities are available, and it is not an inherent advantage of the analytical technique. On the contrary, it can frequently be a disadvantage when the sample to be analysed is heterogeneous. However, for many analytical problems in, for example, environmental, forensic, and clinical applications it is an essential advantage of the technique. The fact should be remembered, though, that sample volumes as small as 50 μl can be nebulized into a conventional flame spectrometer using a microsampling cup technique, with very little reduction in sensitivity compared with continuous flame nebulization.

Many other factors should be considered when comparing electro-thermal atomizers with flame atomization. Electrothermal atomizers, together with the essential fast-response recorder systems, are expensive to purchase, although they can be cheaper to operate than flames. They are, however, safer to operate than flame systems and can be used in enclosed conditions, *e.g.* for the analysis of radioactive materials. Flame atomizers cannot be used for measurements in the vacuum-u.v. whereas inert-gas-purged electrothermal atomizers are easily modified for this work.

Table 2.2   A comparison of atomic absorption absolute detection limits obtained with commercial electrothermal atomizers

Element	Detection limit/pg			
	HGA 74	CRA 63[a]	A3470	IL 355
Ag	0.1	0.2	–	4
Al	2	30	10	90
Au	10	–	–	10
Cd	0.1	0.1	0.6	2
Co	5	4	–	40
Cr	10	2	8	20
Cu	2	4	5	7
Fe	3	4	10	100
Mg	–	0.06	0.2	0.1
Mn	0.2	0.5	3	5
Mo	3	13	20	–
Ni	10	10	10	200
Pb	2	3	3	10
Pt	200	100	–	–
Sn	–	9	200	100
V	100	100	100	400
Zn	0.05	0.08	0.08	1

(a) Figures quoted refer to a tube furnace; detection limits for the cup furnace are normally slightly poorer.

Electrothermal atomizers are more suited for atomic fluorescence measurements than flames. Instrument time per determination is greater for electrothermal atomization than with flame measurements, although the total analysis time can often be shorter owing to the use of a simpler analytical method, *e.g.* a direct solution procedure and determination without prior separation may be possible. Precision for flame measurements (0.2–1 per cent) is better than for electrothermal atomizers (2–5 per cent); however, automatic sampling can improve the precision attainable with electrothermal atomizers.

2.4 Comparison of Detection Limits for the Various Types of Electrothermal Atomizer

A comparison of some atomic absorption detection limits (obtained from the manufacturers' published information), for the atomizers described above, is given in Table 2.2. It should be remembered when comparing the figures that the relative detection limits, which are applicable to solutions, are dependent on the volume of sample that can be used with each atomizer. The detection limits attainable are also dependent on the performance of the manufacturer's atomic absorption spectrometers when coupled with electrothermal atomizers. There is little difference between the absolute detection limits obtained for the three carbon atomizers but the detection limits obtained for the tantalum atomizer are generally poorer. The high sample-volume capability of the graphite furnace does, however, give this type of atomizer superior detection limits when comparing solution concentrations.

3 Theoretical Aspects of the Atomization Process

The atomization processes occurring with electrothermal atomizers would seem to be simpler than the complex interactions occurring with flame atomization. Thermodynamic theories can be postulated which incorporate the dissociation of metal oxides, and other species, and the reduction of metal oxides. However, when the required calculations are carried out for these, and other processes, it becomes apparent that the theoretical results obtained do not always fit readily with results obtained experimentally. Because of this problem, other approaches have concentrated on kinetic theories to obtain information on atomization. While this second approach is probably more helpful analytically, since it can give rise to information that is analytically useful, it is not easy to interpret the experimental results, apart from the simplest cases. It is inevitable therefore that a complete understanding of the situation will come from a combination of both the thermodynamic and kinetic approaches.

An outline of some of the ideas on atomization is given here, but for a thorough treatment the original publications should be consulted.

3.1 Thermodynamic

For a complete understanding of the thermodynamic approach it is necessary to be aware of the various reactions which can occur in an atomizer. To simplify the problem, it will be assumed that the metal salts obtained in an atomizer, after drying the sample solutions, are relatively unstable and therefore are decomposed to their metal oxides prior to any significant atomization occurring. Possible reactions can then be summarised as follows.

(a) Evaporation of the Metal Oxide Prior to Atomization

Some metal oxides have high vapour pressures at the temperatures at which atomization is first observed to occur, *e.g.* Sb_2O_3 and SiO_2. Assuming the simple gas-law equation is valid, it is possible to show that the vapour pressure exerted by, for example, 1 ng of a metal oxide (molecular weight = 100), completely vaporized in a volume of 0.1 ml, at T K, would be 6×10^{-6} T mmHg. If the vapour pressure of the metal oxide is greater than this value, at the minimum atomization temperature, then significant losses of the metal oxide, due to molecular evaporation, would be expected to occur prior to atomization. This problem would be much greater if volatile metal halides were present in the atomizer.

(b) Thermal Dissociation of the Metal Oxide

The degrees of dissociation of metal oxides, α, can be calculated from equations (1) and (2),

$$-\Delta G = RT\ln K_p \tag{1}$$

$$\alpha = K_p / [K_p + \sqrt{P(O_2)}] \tag{2}$$

where $P(O_2)$ is the partial pressure of oxygen, ΔG is the free energy, and K_p is the equilibrium constant for the dissociation of the metal oxide MO as shown in equation (3).

$$MO \rightleftharpoons M + \tfrac{1}{2}O_2 \tag{3}$$

Values for ΔG at various temperatures can be obtained for many metal oxides, from tables of thermodynamic data.

For carbon atomizers, two additional controlling factors are the equilibria existing between oxygen and carbon [equilibria (4) and (5)].

$$2C + O_2 \rightleftharpoons 2CO \tag{4}$$

$$2CO + O_2 \rightleftharpoons 2CO_2 \tag{5}$$

These equilibria effectively control the value of $P(O_2)$.

(c) Reduction of the Metal Oxide

Reduction of the oxide by either the carbon, or the metal from the atomizer is an extension of the theory of dissociation of metal oxides. In this case the dependence of the metal oxide decomposition on the carbon/oxygen equilibria is more apparent. The theory is described here in terms of carbon reduction, although most of the conclusions are equally valid for metal atomizers, *e.g.* tantalum and tungsten.

The reduction process can be described by the equation (6).

$$MO(s/l) + C(s) \rightleftharpoons M(g) + CO(g) \tag{6}$$

The free-energy change for the reaction, ΔG°, can be obtained from the difference in the sum of the free energies of formation of the products and reactants. In this calculation it is essential to use ΔG values which correspond to the correct physical state of each component, as described by Campbell and Ottaway[34] (Table 3.1). By using the various free-energy values for reactants and products over a range of temperatures it is possible to calculate the temperature at which formation of the free gaseous metal becomes feasible, *i.e.* ΔG° becomes negative.

It is sometimes easier to use Ellingham diagrams[35] for these calculations, where the thermodynamic information is presented graphically (Figure 3-1). These diagrams were originally used for studying metallurgical reduction processes, so they require modifications to incorporate the added criterion for the formation of gaseous metal atoms.

The difficulty in using this thermodynamic model for atomization is in relating theoretically derived values with results obtained experimentally.

Table 3.1 Calculation[34] of the free energy of reaction for
$Al_2O_3(s) + 3C(s) \rightarrow 2Al(g) + 3CO(g)$

Temperature/ K	ΔG_f°/kcal mol^{-1}				ΔG° (reaction)/ kcal mol^{-1}
	Products		Reactants		
	2Al(g)	3CO(g)	Al_2O_3(s)	3C(s)	
2300	+23	−223	−225	0	+25
2400	+17	−229	−218	0	+6
2500	+12	−235	−212	0	−11
2600	+7	−241	−206	0	−28

This problem arises for two reasons. (i) The thermodynamic calculation of
the minimum atomization temperature $(\Delta G^{\circ} = 0)$ is dependent on an
equilibrium reaction; therefore some proportion of the metal oxide will be
dissociated at *all* temperatures. The quantity of metal atoms formed will
be dependent on the quantity of metal oxide present in the atomizer, *i.e.*

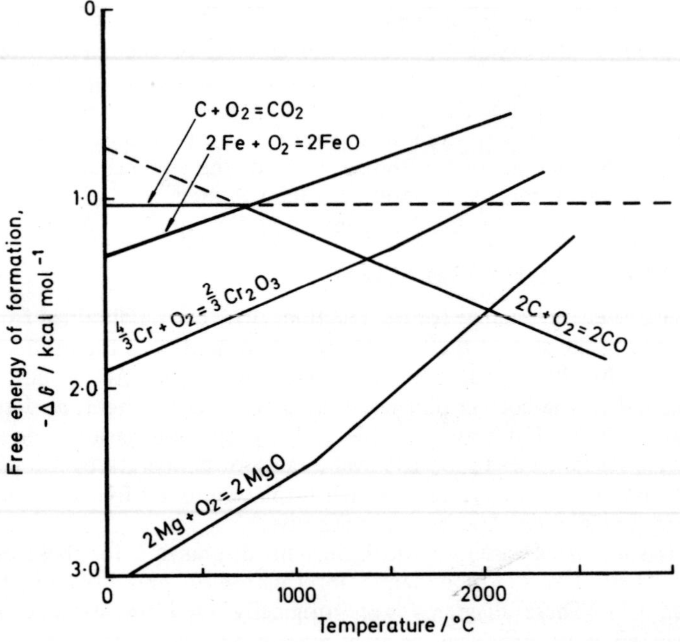

Figure 3-1   An Ellingham diagram for the reduction of metal oxides by carbon. A
reduction process becomes favourable at the temperature where the free energy
of formation of a metal oxide becomes equal to that of carbon dioxide or
carbon monoxide.

the greater the quantity of metal oxide present the greater the concentration of metal atoms formed at any temperature. (ii) The minimum concentration of an element which can be detected is dependent on the detection limit (compare silver and vanadium in a carbon furnace by atomic absorption at 0.1 and 100 pg, respectively) and also on the analytical technique (compare the detection limits by atomic fluorescence and atomic absorption for gold at 5 and 200 pg respectively[36] and for manganese at 1 and 50 pg respectively,[37] using a carbon rod atomizer). Quite different analytical conditions could therefore be used, with the result that widely differing values are obtained for the minimum temperatures at which atomization is observed. It is in fact theoretically correct to say that any temperature could be obtained since the value obtained is solely dependent on the conditions chosen (albeit highly improbable).

Campbell and Ottaway[34] (carbon furnace) and Aggett and Sprott[38] (carbon rod) have used the carbon-reduction model and compared experimental and theoretical values of the minimum atomization temperatures. Two other groups of workers[39,40] have measured minimum atomization temperatures for a carbon cup and a carbon rod. The results obtained by these four groups of workers are summarized in Table 3.2.

While the results obtained with the furnace agree with the theoretical values in most cases, it is interesting to note that in no case do all four atomizer temperatures agree. Even with the two sets of results for the carbon rod, three out of seven temperatures vary by 150 °C or greater. It would be extremely difficult to draw any conclusions from these results

Table 3.2 Comparison of theoretically predicted minimum atomization temperatures and experimental values obtained using carbon furnace, carbon rod, and carbon cup atomizers.

Element (as oxide)	Minimum atomization temperature/K				
	Theoretical[34]	Furnace[34]	Rod[38]	Rod[39]	Cup[40]
Al_2O_3	2450	2300	1820	–	1860
CdO	>800	850	540	710	570
CaO	2450	1800	1820	–	1660
Cr_2O_3	1850	1800	1650	–	1280
CoO	1800	1720	1550	1700	1220
Cu_2O	1850	1730	1430	–	1050
Fe_3O_4	1750	1750	1500	–	1150
PbO	1050	1000	1130	–	780
MgO	2150	1550	1570	1530	1200
Mn_3O_4	1650	1600	1480	1470	1270
NiO	1750	1800	1690	1620	1310
Ag_2O	1250	1150	1120	1110	780
SnO_2	1350	1800	1410	–	1300
ZnO	1200	1100	1070	870	730

other than that the furnace appears to fit this atomization model whereas the rod and cup do not. A comparison of minimum atomization temperatures for furnace, rod, and cup atomizers should, however, be conducted under exactly comparable conditions (*i.e.* concentration, temperature measurement, and general criteria for defining minimum detectable atomization). These temperatures would then show whether any significant differences are obtained or whether the differences obtained above are only a function of the different analytical conditions used by the four groups of workers.

(d) Carbide Formation

In the same way that free-energy changes can be calculated for carbon reduction of metal oxides it is possible to calculate the free-energy changes occurring in the formation of stable carbide compounds:

$$MO + 2C \rightleftharpoons MC + CO \tag{7}$$

Carrying out this calculation frequently shows that the metal oxide decomposes to form a stable carbide at temperatures below that at which carbon reduction of the metal oxide to gaseous metal atoms occurs, *e.g.* Al, Ca, and Cr. This temperature is also often below that at which atomization is first observed.

(e) Analytical Applications of Thermodynamic Theories of Atomization

Very little work has been published on the application of thermodynamic data to analytical problems. This is surprising, considering that it should be possible to predict many interference effects arising from the formation of stable or unstable compounds.

Campbell[41] has attempted to explain the differences in performance between carbon and tantalum atomizers by calculating the theoretical temperatures at which atomization should first be observed for both carbon and tantalum reduction of metal oxides [see Section 3.1 (c)]. Due to the increasing stability of carbon monoxide and the decreasing stability of tantalum oxide with increasing temperature, the calculations produced the following conclusions: (i) when the temperature for $\Delta G^{\circ} = 0$ is about 1600 K there is little difference between carbon and tantalum, (ii) when the temperature for $\Delta G^{\circ} = 0$ is less than 1600 K for carbon then the corresponding temperature for tantalum is even lower, and (iii) when the temperature for $\Delta G^{\circ} = 0$ is greater than 1600 K for carbon the temperature for tantalum is higher. Therefore it can be predicted that elements readily atomized on carbon (*e.g.* cadmium and lead) are even more readily atomized on tantalum, and elements which are atomized with difficulty on carbon (*e.g.* silicon and titanium) are very difficult to atomize on tantalum. While these results do fit with general observations, as yet there are no comprehensive experimental results available to confirm the theory.

Frech and Cedergren[41a] have used high-temperature equilibrium calculations to investigate the interference effect of chlorine on the determination of lead in steel samples. Using a computer program, they considered the effects of 9 elements (Fe, Cl, Pb, S, H, O, N, Ar, and C) present as 36 different compounds. From the calculations it was possible to predict that a sample should be ashed at 900 K in the presence of hydrogen to minimize interferences. If other systems are studied in a similar way this approach shows considerable promise as a means of understanding and predicting interference effects in general.

From the previous sections, it is apparent that there are five important thermodynamic parameters. (i) H_{MO}, the heat of vaporization of the metal oxide, (ii) E_{MO}, the dissociation energy of the metal oxide, (iii) H_M, the heat of vaporization of the free metal, (iv) E_{MC}, the dissociation energy of the metal carbide, and (v) E_{CO}, the dissociation energy of carbon monoxide (which is common for all elements).

Several situations can be predicted, therefore. (i) If $H_{MO} < E_{MO}$ and E_{MC}, vaporization of the metal oxide will occur prior to any reaction. (ii) If E_{MO} and $E_{MC} < H_M$ then reduction of the metal oxide or carbide will occur rapidly at the atomization temperature, and the rate of atomization will be dependent on H_M. (iii) If $H_M < E_{MO} < E_{MC}$ the rate of atomization will be dependent on E_{MC}. (iv) If $H_M < E_{MC} < E_{MO}$ the rate of atomization will be dependent on E_{MO}.

While thermodynamic considerations are clearly important in the atomization processes, it is apparent that these processes must also be kinetically controlled to explain the experimental results.

3.2 Kinetic

The kinetic models which have been proposed vary from the simple to the complex; unfortunately, no more than a brief description of each can be given here. The models studied fall into two groups, (i) atomization under an increasing temperature, first used by L'vov,[42] and (ii) atomization under isothermal conditions, proposed by Fuller.[43] Neither of these models is entirely satisfactory although each has advantages. L'vov's model approximates the situation with rod, cup, and filament atomizers and probably for easily atomized elements at high temperatures in a furnace. Fuller's model approaches the situation existing in furnace atomizers, particularly for elements which are difficult to atomize and at low temperatures. Both models can be applied to the pyrolysis stage of an analytical determination; L'vov's for ramp pyrolysis and Fuller's for isothermal pyrolysis.

(a) Atomization under increasing Temperature
L'vov proposed equation (8) for atomization in a furnace,[42] where $\mathrm{d}N/\mathrm{d}t$ is the rate of change in the number of atoms, N, present in the gaseous state in the atomizer, $n_1(t)$ is the number of atoms entering the system,

and $n_2(t)$ is the number of atoms leaving the system. For a constantly increasing atomization temperature, equations (9) and (10) are valid, where τ_1 is the time taken to transfer the total number of atoms, $N(0)$, to the system; hence one can obtain equation (11).

$$dN/dt = n_1(t) - n_2(t) \tag{8}$$

$$n_1(t) = At \tag{9}$$

$$\int_{t=0}^{t=\tau_1} n_1(t)\,dt = N(0) \tag{10}$$

$$n_1(t) = 2N(0)\,t/\tau_1^2 \tag{11}$$

Assuming that atoms are removed from the system by vapour diffusion, then equation (12) applies,

$$n_2(t) = N/\tau_2 \tag{12}$$

where τ_2 is the average residence time of atoms in the system. By substituting equations (11) and (12) into equation (8) one may obtain equation (13).

$$dN/dt = [2N(0)t/\tau_1^2] - (N/\tau_2) \tag{13}$$

L'vov describes two situations when integrating equation (13).
For $t \leqslant \tau_1$,

$$N = 2N(0)\tau_2^2\,[(t/\tau_2) - 1 + \exp(-t/\tau_2)]/\tau_1^2 \tag{14}$$

and $t \geqslant \tau_2$

$$N = 2N(0)\tau_2^2\{[(\tau_1/\tau_2) - 1 + \exp(-t/\tau_2)]\,\exp[(\tau_1 - t)/\tau_2]\}/\tau_1^2 \tag{15}$$

Equations (14) and (15) together describe the number of atoms, N, in the system at time t.

A second approach to this model has been made by Torsi and Tessari[44–47] for a carbon rod atomizer. These authors assumed that the space around the carbon rod acted as an infinite sink for the evaporated atoms and that the sample was deposited on the rod as a monatomic layer.

The rate of release of atoms from the surface, $n_1(t)$, is given by equation (16),

$$n_1(t) = -sq\,d\theta/dt = ksq\theta \tag{16}$$

where s is the surface area/cm^2, θ is the fraction of the surface covered at time $t(0 \leqslant \theta \leqslant 1)$, q/atoms cm^{-2} is the surface concentration when $\theta = 1$, and k/s^{-1} is the rate constant for the evaporation process.

If the rate constant, k, shows the normal Arrhenius temperature dependence then equation (16) can be rewritten in the form (17).

$$n_1(t) = Aqs\theta\,\exp(-\Delta G/RT) \tag{17}$$

This equation describes a typical peak signal, because as T increases with time there is a corresponding reduction in the value of θ as atoms are lost from the rod surface. Assuming that the temperature varies with time as the function $T = T(0) + \alpha T$, where $T(0)$ is the initial temperature and $\alpha = \mathrm{d}T/\mathrm{d}t$ (a constant), the value of $n_1(t)$ varies according to equation (18).

$$n_1(t) = Aqs\,\theta\,\exp\{-\Delta G/R\,[T(0) + \alpha t]\} \tag{18}$$

This model was refined in the later work[45,46] of these authors in terms of atom-transport effects and thermal perturbation of the atomizer.

A very similar approach to this model has been made by Johnson, Sharp, West, and Dagnall.[39] Vaporization from a solid or liquid surface can only occur when atoms acquire a minimum energy ϵ; and the number of atoms, $n(\epsilon)$, in a system with an energy greater than or equal to ϵ is proportional to $\exp(-\epsilon/BT)$, where B is the Boltzmann constant, *i.e.* the gas constant per atom or molecule. If a monatomic layer of atoms, N, exists on the carbon rod then:

$$n(\epsilon) = N\exp(-E/RT) \tag{19}$$

Since the rate of vaporization will be proportional to this number, equation (20) may be derived,

$$-\,\mathrm{d}N/\mathrm{d}t = CN\exp(-E/RT) \tag{20}$$

where C is a constant dependent on the transfer of energy from the atomizer to the sample. Since T varies with time $[T = \mathrm{f}(t)]$, equation (20) can be integrated to obtain the number of atoms, $N(t)$, remaining on the rod after time t [equation (21)].

$$N(t) = N(0)\exp\{-\,C\int_{t=0}^{t=t}\exp[-E/R\mathrm{f}(t)]\;\mathrm{d}t\} \tag{21}$$

Assuming that the analysis volume is directly above the rod, that removal of atoms from this volume occurs at the same rate as they enter, and that the residence time of atoms in this volume is τ, then it is possible to determine the number of atoms, $A(t)$, which are present in the analysis volume by calculating the difference between the number of atoms which have entered and left the volume at time t, *i.e.*

$$A(t) = [N(0) - N(t)] - [N(0) - N(t - \tau)] = N(t - \tau) - N(t) \tag{22}$$

Using equation (21),

$$A(t) = N(0)\,\{\exp(-\,C\int_{t=0}^{t=t}\exp[-E/R\mathrm{f}(t)]\,\mathrm{d}t)$$

$$-\exp\,(-\,C\int_{t=0}^{t=t-\tau}\exp[-E/R\mathrm{f}(t)]\;\mathrm{d}t)\} \tag{23}$$

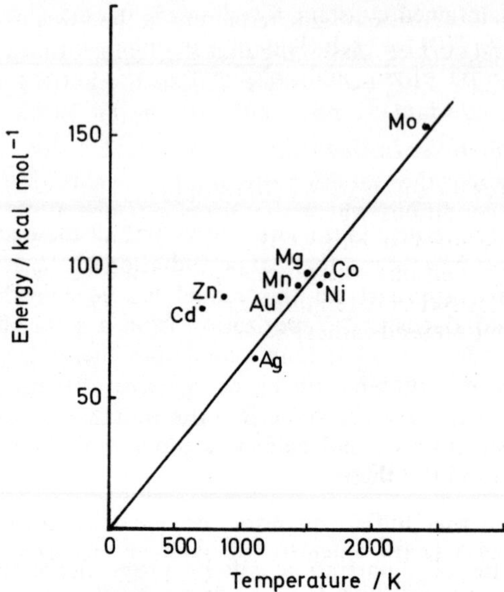

Figure 3-2   The variation of the energy, relating to atomization, as a function of the temperature at which atomic absorption is first observed for a carbon rod atomizer. H_M: Ag, Au, and Mo; E_{MO}: Cd, Zn, Mn, Mg, Ni, and Co.

Values of τ and f(t) can be obtained experimentally, and E represents the minimum energy per mole required for an atom to leave the surface of the carbon rod. E can be related to three parameters: E_{MO}, the bond energy of the metal oxide; H_M, the heat of vaporization of the element; and H_{MO}, the heat of vaporization of the metal oxide. Section 3.1 summarizes the dependence of the atomization process on these thermodynamic quantities.

Agreement between experimental absorbance profiles and results predicted by equation (23) was not generally very good. However, since the residence time τ (*ca.* 0.01 seconds) is small compared with the atomization time t (*ca.* 0.2–1 seconds), $A(t)$ is mainly governed by equation (20). At the temperature when atomization just commences, $T(0)$, the number of atoms remaining on the rod is still approximately $N(0)$ and the number of atoms in the analysis volume is $A(0)$, so that:

$$A(0) \propto CN(0) \exp[-E/RT(0)] \tag{24}$$

At low absorption values the observed signal, S, is proportional to Af, where f is the oscillator strength; so if $N(0)$ is made proportional to $1/f$ for each element then

$$S(0) \propto C \exp[-E/RT(0)] \tag{25}$$

If $S(0)$ is maintained constant for all elements and C is assumed to be constant, then $E/T(0)$ for each element should be constant; see Figure 3-2. With the exception of zinc and cadmium, good agreement was obtained.

(b) Atomization under Isothermal Conditions

Fuller[43] measured the variation in atom concentration in a graphite furnace under conditions where the furnace temperature was essentially constant for the greatest part of the atomization process. This entailed working under conditions which were optimized for kinetic measurements rather than for analytical determinations.

With the notation used earlier,

$$dN/dt = n_1(t) - n_2(t) \tag{8}$$

here,

$$n_1(t) = k_1 [N(0) - N(t)] \tag{26}$$

where $N(0)$ is the initial quantity of element introduced to the atomizer and $N(t)$ is the quantity of element atomized up to time t, and k_1 is the rate constant for the atomization process. $N(t)$ can be calculated from equation (27):

$$N(t) = N(0) [1 - \exp(-k_1 t)] \tag{27}$$

which on substitution into equation (26) gives equation (28).

$$n_1(t) = k_1 N(0) \exp(-k_1 t) \tag{28}$$

The rate of loss of atoms from the furnace is controlled by diffusion processes and the velocity of the purge gas passing through the furnace, hence

$$n_2(t) = k_2 N \tag{29}$$

where k_2 is the rate constant for the removal of atoms from the furnace. Substituting equations (28) and (29) into equation (8), one obtains equation (30), which on integration produces equation (31), representing the concentration of atoms at any time.

$$dN/dt = k_1 N(0)\exp(-k_1 t) - k_2 N \tag{30}$$

$$N = [k_1/(k_2 - k_1)]N(0)[\exp(-k_1 t) - \exp(-k_2 t)] \tag{31}$$

(c) Analytical Applications of Kinetic Theories of Atomization

The main advantage of the kinetic models of atomization is that they are used to produce analytically useful information. This will be demonstrated, using the theory for isothermal atomization in a graphite furnace, although similar information could be derived from other theories.

(i) Calibration by Signal Peak Height and Signal Integration.[48] Equation (31) can be differentiated to obtain the peak absorbance signal [equation (32)] and integrated to obtain the integrated signal [equation (33)].

$$N_{\text{peak}} = N(0) \left(\frac{k_2}{k_1}\right)^{k_2/(k_1 - k_2)} \tag{32}$$

$$[N_{\text{integrated}}]_{t=0}^{t=t} = N(0) \frac{k_1}{k_2 - k_1} \left[\frac{1}{k_2}\exp(-k_2 t) - \frac{1}{k_1}\exp(-k_1 t)\right]_{t=0}^{t=t} \tag{33}$$

The ratios of N_{peak} to $N_{\text{integrated}}$ can be compared to obtain the conditions where the measurement of the integrated signal is advantageous compared to peak height measurement. The ratio tends to 1 at high temperatures (*i.e.* $k_1 \gg k_2$), where both signals become equal to $N(0)$, and to zero at low temperatures (*i.e.* $k_1 \ll k_2$), where N_{peak} tends to zero. Therefore integration is preferable for measurements at low temperatures (relative) or low atomization rates.

(ii) Stopped Gas Flow.[48] Graphite furnaces can be operated under conditions of stopped gas flow during the atomization stage to obtain increased sensitivity for many elements.[24] The idealized situation for stopped-gas-flow operation is given by setting $k_2 = 0$ in equation (30), which is effectively a simple signal-integration technique. The maximum signal enhancement which can be obtained is given by the ratio of $N(0)$ to equation (32), *i.e.*

$$\text{Maximum signal enhancement} = \left(\frac{k_2}{k_1}\right)^{-k_2/(k_1 - k_2)} \tag{34}$$

In practice this situation is not achieved, because the component of k_2 corresponding to removal of atoms by diffusion cannot be reduced to zero.

(iii) Matrix Control. It is possible to predict whether the determination of an element A in the presence of a matrix B is feasible and to define the optimum analytical conditions. For an analytical determination it is required that the concentration of the matrix element vaporized during the atomization stage be below that at which it will interfere in the determination. Assuming that less than 100 μg ml^{-1} of the matrix does not interfere and that the concentration of the matrix in the sample solution is 10 000 μg ml^{-1} then it is necessary to ensure that no more than 1 per cent of the matrix is vaporized during the atomization stage.

The losses of elements during the pyrolysis stage of a determination car

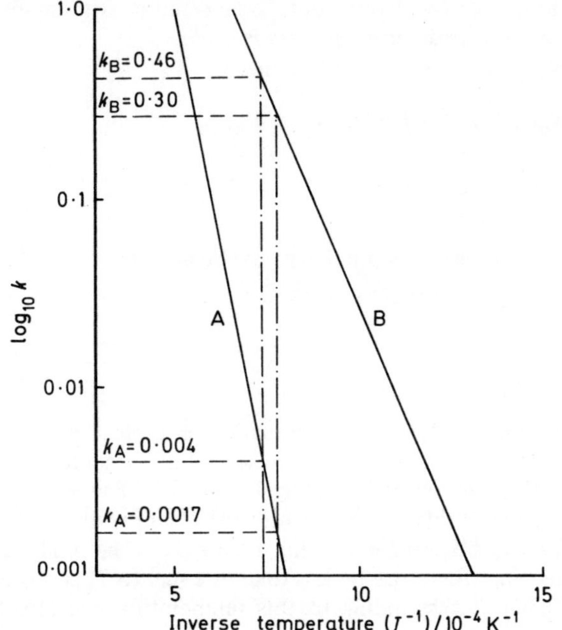

Figure 3-3   Arrhenius temperature dependence of the rate constants.

be obtained over a range of pyrolysis temperatures by plotting the atomization signals obtained as a function of pyrolysis time.[49] The kinetics for the loss of each element from the furnace can then be derived from this information. The variation of $\log_{10} k$ with $1/(T/\mathrm{K})$ can be plotted for each element on to a single graph, Figure 3-3, and the results extrapolated to obtain values of k at all values of T.

Assuming that the loss of elements from the furnace obeys first-order kinetics, that C represents the concentration of the element remaining in the furnace at time t, and that $C(0)$ represents the initial concentration of the element in the furnace at time $t = 0$ then

$$-\mathrm{d}C/\mathrm{d}t = kC \tag{35}$$

which on integration yields

$$\ln[C(0)/C] = kt \tag{36}$$

The first analytical situation is where the matrix (B) is more readily vaporized than the analyte element (A). Here, it is necessary to remove 99 per cent of the sample matrix prior to the atomization of the analyte. However, it is essential that the concentration of the analyte is not decreased significantly, *i.e.* by greater than the precision of a routine determination, *e.g.* by 5 per cent. Using these criteria, one can now establish whether the determination is feasible.

For removal of 99 per cent of the matrix

$$\ln\{\ [B(0)] \times 100/[B(0) \times 1\ \} = k_B t_{99} \tag{37}$$

so,

$$t_{99} = 4.6/k_B \tag{38}$$

For the loss of less than 5 per cent of the analyte

$$\ln\{[A(0)] \times 100/[A(0)] \times 95\} = k_A t_5 \tag{39}$$

so,

$$t_5 = 0.051/k_A \tag{40}$$

For a normal analytical determination a typical pyrolysis period is 30–60 seconds; therefore for element A and a 30 second pyrolysis period $k_A = 0.0017$ can be obtained, using Figure 3-3. The analytical pyrolysis outlined). From this information the temperature, T, at which $k_A = 0.0017$ can be obtained, using Figure 3.3. The analytical pyrolysis temperature must therefore be less than or equal to T, and the value of k_B (equal to 0.30) corresponding to this temperature can also be obtained using Figure 3-3. This value of k_B can be substituted into equation (38) to obtain a value of $t = 15$ s; therefore, for a pyrolysis period of 30 seconds, in excess of 99 per cent of the matrix will be removed. This procedure can be simplified by plotting a graph of percentage loss of matrix against k_B for a 30 second pyrolysis period, using equation (37); Figure 3-4. It is

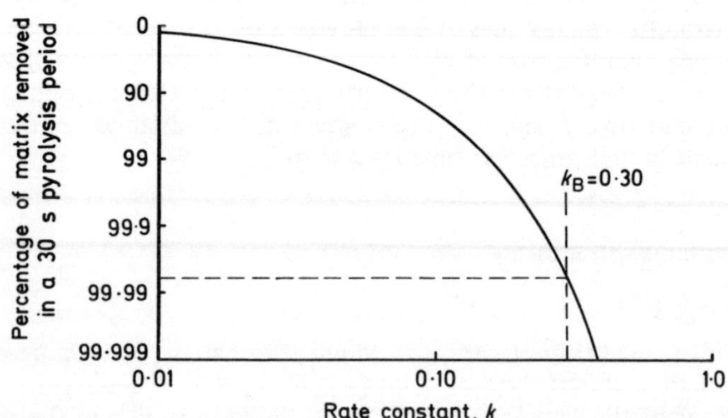

Figure 3-4   The variation in the percentage loss of matrix with the rate constant (pyrolysis period is 30 seconds).

now only necessary to use the value of $k_B = 0.30$ in this graph to see that greater than 99.9 per cent of the matrix is removed.

The second analytical situation is where the analyte element (now B) is more readily vaporized than the matrix (now A). Here the criteria are reversed, and it is necessary to atomize up to 90 per cent of the analyte before 1 per cent of the matrix is vaporized. For the removal of 90 per cent of the analyte,

$$\ln\{[B(0)] \times 100/[B(0)] \times 10\} = k_B t_{90} \tag{41}$$

$$t_{90} = 2.3/k_B \tag{42}$$

For the removal of 1 per cent of the matrix,

$$\ln\{[A(0)] \times 100/[A(0)] \times 99\} = k_A t_1 \tag{43}$$

$$t_1 = 0.0092/k_A \tag{44}$$

To obtain adequate sensitivity the atomization period should not exceed 5 seconds; therefore from equation (42), $k_B = 0.46$ (this, again, is a constant for all elements under the conditions outlined). Using Figure 3-3, the value of k_A (equal to 0.004), corresponding to the temperature at which $k_B = 0.46$, can be obtained. The percentage of the sample matrix vaporized under these conditions can be obtained from Figure 3-5, constructed from equation (43). Since greater than 1 per cent of the matrix is vaporized this determination would not be possible without experiencing the interference effect.

However, this method of assessing analytical feasibility is obviously grossly simplified, since a number of assumptions have been made which

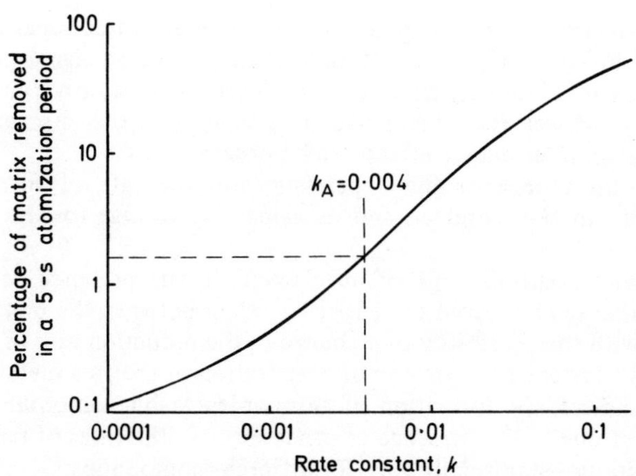

Figure 3-5   The variation in the percentage loss of matrix with the rate constant (atomization period is 5 seconds).

may or may not be true in practice, *e.g.* a simple first-order kinetic equation is used, the kinetics for loss of elements at low temperatures are the same at high temperatures, the kinetics for low concentrations of elements are the same at high concentrations, and there are no variations in kinetic behaviour from one sample medium to another. The model could, however, be refined to take into account many of these variations.

(iv) Interference Studies.[49a] Interferences in the analytical applications of electrothermal atomizers have been seriously neglected. Kinetic studies of the atomization and pyrolysis stages of determinations would provide an ideal method of investigating these interference effects. For example, kinetic studies of the atomization profile could provide information on whether differences in atomization signals are caused solely by variations in the rate of release of atoms or whether there is a real variation in the number of atoms produced. Also, kinetic studies of the pyrolysis stage could show whether atomization signal variations are caused by losses of atomic or molecular species prior to atomization.

Assume that a general equation (30), similar to that derived for copper,[43] can be written to describe the concentration of metal atoms in a graphite furnace. By inserting a proportionality factor, p, into the integrated equation (31), it is possible to obtain an equation (45) which gives the measured absorbance at any time, t. This factor p will be a function of the oscillator strength for each element and the efficiency with which metal atoms are produced.

$$\text{Absorbance } (N) = pN(0)\,[k_1/(k_2 - k_1)]\,[\exp(-k_1 t) - \exp(-k_2 t)]$$

$$(45)$$

From equation (45) it is apparent that there are three separate factors which will influence the shape of an atomic absorption signal profile, *i.e.* k_1, k_2, and p. If, during an analytical determination, any one of these parameters changes then there will be a change in the absorbance/time profile, and an interference effect will be observed.

k_1: this represents the dependence on the rate of formation of metal atoms in the atomizer, and its value may change for the following reasons:

(i) Physical: entrainment of an element in the presence of a matrix causing either (*a*) increased or reduced contact between the metal salt and graphite, with the possibility of a change in the reduction rate of the metal salt, or (*b*) a faster or slower rate of evaporation of the free metal.

(ii) Chemical: the formation of more or less stable compounds prior to atomization due to the presence of other species. The rates of reduction or decomposition will normally vary for different compounds.

Figure 3-6 shows that for variations in the value of k_1 (0.01–1.0 s^{-1}) at fixed values of k_2 (0.1 s^{-1}), p (1 x 10^8), and $N(0)$ (1 x 10^{-8} g) the

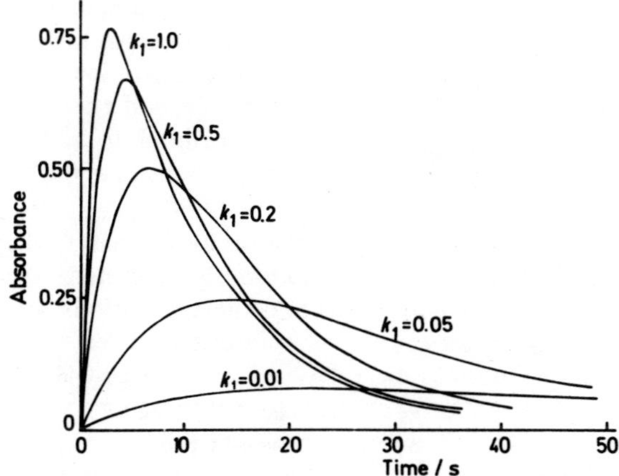

Figure 3-6   The effect of variations in the value of k_1 on an absorbance–time profile. k_2 = 0.1 s^{-1}, p = 1 × 10^8, $N(0)$ = 1 × 10^{-8} g.

peak height value increases and the time to reach the maximum absorbance value decreases with increasing values of k_1.

k_2: this represents the dependence on the rate of removal of metal atoms from the atomizer, and its value may change for the following reasons:

(i) A change in the flow rate of the inert gas through the atomizer.

(ii) A change in the diffusion rate of metal atoms through the walls of the graphite furnace caused by (*a*) the degradation of the graphite furnace or (*b*) the formation of a pyrolytic graphite coating on the graphite furnace.

Figure 3-7 shows that for variations in the value of k_2 (0.01–1.0 s^{-1}) at fixed values of k_1 (0.1 s^{-1}), p (1 × 10^8), and $N(0)$ (1 × 10^{-8} g) the peak height value and the time to reach the maximum absorbance value increase with decreasing values of k_2.

p: this represents the dependence on the proportion of the element, introduced to the atomizer, which is converted into vaporized metal atoms compared to the vaporization of other species, *e.g.* oxides, chlorides, and organometallics. It therefore reflects any changes in the atomization efficiency.

Figure 3-8 shows that for variations in the value of p (0.1–4.0 × 10^8) at fixed values of k_1 (0.1 s^{-1}), k_2 (0.2 s^{-1}), and $N(0)$ (1 × 10^{-8} g) the peak height value increases with increasing values of p. The time to reach the maximum absorbance value is independent of changes in the value of p.

Further information can be obtained by looking at the variation of the integrated signal with variations in the values of k_1, k_2, and p.

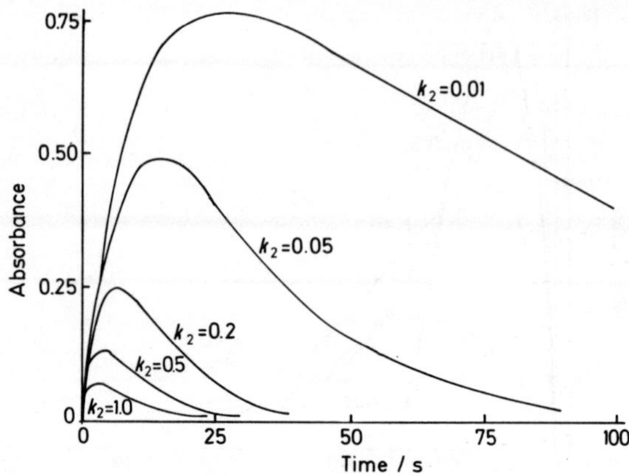

Figure 3-7   The effect of variations in the value of k_2 on an absorbance—time profile. $k_1 = 0.1$ s^{-1}, $p = 1 \times 10^8$, $N(0) = 1 \times 10^{-8}$ g.

From equation (45),

$$[\text{Absorbance}]_{t=0}^{t=\infty} = pN(0)[k_1/(k_2 - k_1)] \times$$

$$\left[\frac{\exp(-k_2 t)}{k_2} - \frac{\exp(-k_1 t)}{k_1}\right]_{t=0}^{t=\infty} \tag{46}$$

$$[\text{Absorbance}]_{t=0}^{t=\infty} = pN(0)/k_2 \tag{47}$$

The integrated absorbance signal is therefore independent of k_1 but linearly proportional to p and inversely proportional to k_2. Equation (47) shows clearly why signal-integration techniques cannot compensate for interference effects in all cases. Compensation is only possible for one of the three causes of interference, *i.e.* where the rate of atomization (k_1) varies.

The information, described above, for assessing the causes of interference effects is summarized in Table 3.3. This summary shows that the change in any one parameter is not a positive guide to the cause of an interference effect. However, by comparing the changes in all three parameters, an immediate guide to the type and cause of interference can be obtained, even where two separate interference effects occur simultaneously.

(v) The Determination of Reaction Mechanisms. As with all kinetic measurements, it is possible to postulate reaction mechanisms that are consistent with the data obtained. However, there are many possible reactions which would fit simple kinetics for the formation of atoms in a

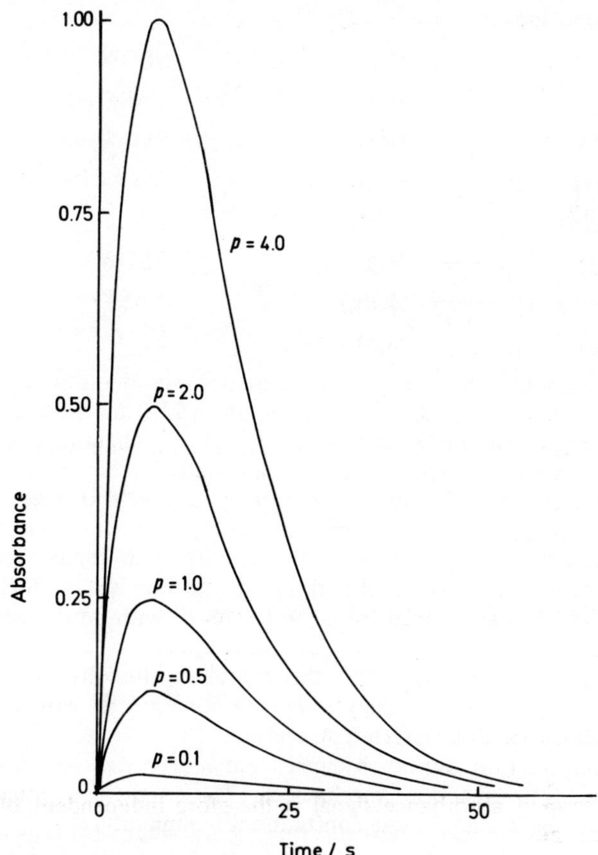

Figure 3-8   The effect of variations in the value of p on an absorbance–time profile. $k_1 = 0.1$ s^{-1}, $k_2 = 0.2$ s^{-1}, $N(0) = 1 \times 10^{-8}$ g.

Table 3.3   A summary of the effects, on peak absorbance, time to reach peak absorbance, and integrated absorbance, caused by *increasing* the values of the three parameters k_1, k_2, and p.

Increased variable	Effect on Observed Parameter		
	Peak absorbance	Time to reach peak absorbance	Integrated absorbance
k_1	increase	decrease	constant
k_2	decrease	decrease	decrease
p	increase	constant	increase

graphite atomizer; *e.g.*

$$MO(s/l) \longrightarrow M(g) + \tfrac{1}{2}O_2 \qquad \text{SLOW} \qquad (48)$$

$$MO(s/l) \longrightarrow M(s/l) + \tfrac{1}{2}O_2 \qquad \text{FAST} \qquad (49a)$$

$$M(s/l) \longrightarrow M(g) \qquad \text{SLOW} \qquad (49b)$$

$$MO(s/l) + C \longrightarrow M(g) + CO \qquad \text{SLOW} \qquad (50)$$

$$MO(s/l) + C \longrightarrow M(s/l) + CO \qquad \text{FAST} \qquad (51a)$$

$$M(s/l) \longrightarrow M(g) \qquad \text{SLOW} \qquad (51b)$$

$$MO(s/l) \longrightarrow MO(g) \qquad \text{FAST} \qquad (52a)$$

$$MO(g) \longrightarrow M(g) + \tfrac{1}{2}O_2 \qquad \text{SLOW} \qquad (52b)$$

These examples show that atom formation could occur through the dissociation of metal oxides [(48), (49), and (52)] or the reduction of metal oxides by carbon [(50) and (51)]. The mechanism expressed in equation (52) would explain the experimentally observed fact that the efficiency of atomization can vary with temperature.[43] It is clear that the rate of vaporization of metal oxide molecules and the rate of dissociation of the metal oxide molecules would not vary with temperature in an identical manner. Therefore the ratio of the number of metal oxide molecules which are removed from the atomizer, without dissociation, to the number of metal oxide molecules which are dissociated will vary with temperature. On a practical level, this would explain why the integrated absorbance signal in an analytical determination is not necessarily independent of the atomization temperature.

Other information can be obtained from kinetic data which may help to elucidate the reaction mechanism. The Arrhenius equation (53) expresses the variation of a rate constant with temperature.

$$k = A \exp(-E/RT) \qquad (53)$$

By plotting a graph of $\ln k_1$ against $1/T$, a linear relationship should be obtained, where the slope of the graph is equal to $-E/R$. Since the value of E represents the energy required, in the rate-determining step, to produce metal atoms, its value may be related to the thermodynamic functions described in Section 3.1.

4 General Experimental Conditions

4.1 Sample Handling

The increased sensitivity associated with electrothermal atomization may be capable of solving many of the analytical chemist's problems but it could also introduce him to a new range of problems connected with sample handling. Unfortunately, insufficient emphasis has been placed on these problems in the literature. For true ultra-trace analysis the importance of clean-air rooms, or at worst clean areas, cannot be overstated; particularly for the determination of some of the common contaminating elements, *e.g.* Fe, Cu, Ni, Cr, and Na, and frequently for many of the less obvious elements, *e.g.* Cd, Pb, and Zn.

Space available here does not permit a lengthy description of all the problems encountered or the means to overcome them. However, Mitchell[50] has clearly defined many of the problems existing with this type of work; including among them (*a*) contamination from atmospheric dust and materials of construction within the laboratory, (*b*) preparation of ultrapure water, acids, and other reagents for sample preparation, (*c*) storage of pure reagents, and (*d*) routine operation under clean conditions. While it is frequently uneconomic to build clean-air rooms and to purify commercially available acids and reagents for a limited amount of work, it is essential (for many of the analytical results obtained to be meaningful) to appreciate the problems involved.

By careful laboratory management and a minimum of expenditure it is possible to convert an existing small laboratory into one operating under clean conditions, with the outcome that acceptable analytical results can be obtained for most applications. Zief and Nesher[51] have suggested the following modifications for a laboratory (*a*) all incoming air sources, from air conditioning, heating, and ventilation, should be fitted with high-efficiency filters, (*b*) all walls and ceilings should be coated with a non-shedding enamel paint, (*c*) floors should have a vinyl covering, (*d*) unnecessary shelving, partitions, and furniture should be removed, (*e*) windows and wall openings (apart from the door) should be made airtight, and (*f*) the room should be kept under constant positive air pressure at least 1.5 mmHg above ambient. They also suggested that clean-air modules should be used in the critical work areas. To these suggestions could be added the use of the minimum quantity of laboratory apparatus (beakers, volumetric flasks, pipettes, *etc.*), which should be retained solely for this type of work, and the use of the highest purity chemicals available which are compatible with the requirements of the analysis.

Having outlined briefly the general problems involved with handling, it is now appropriate to detail some of the more specific problems encountered in analysing solid, liquid, and gaseous samples.

(a) Solids

The direct determination of trace impurities in solids is highly desirable since it can eliminate time-consuming sample decompositions and problems of reagent purity (leading to high sample blanks). The capability of electrothermal atomization to carry out this type of analysis is a distinct advantage of the technique over conventional flame spectrometry.

Direct solids analysis has been successfully achieved using the cup and furnace configurations. The cup design is ideally suited to solids analysis since a sample can be weighed directly into the cup prior to its insertion into the atomizer work head. With the furnace, solid samples are normally weighed into a small tantalum boat, which is inserted into the furnace using a specially designed tool; the boat can then be inverted so that the sample is deposited reproducibly in the centre of the furnace. For atmospheric particulates analysis see Section 4.1.c.

The critical factor in the determination of trace elements in solids without a prior separation (whether it is by direct solids analysis or by a simple solution procedure) is the limiting ratio of the trace element to matrix quantity. Clearly, one cannot go on increasing the weight of sample introduced to the furnace in order to obtain lower detection limits. In practice $0.1 \, \mu\mathrm{g}\,\mathrm{g}^{-1}$ is a reasonable level which one can expect to achieve, although the actual figure will be dependent on the sensitivity of the element being determined. For samples consisting of a volatile or easily pyrolysed matrix (*e.g.* organic compounds) or a matrix which can be converted into a volatile form (*e.g.* silica to silicon tetrafluoride) larger samples can be introduced to the atomizer, since the matrix will be removed prior to atomization, during the pyrolysis stage. In this way lower detection limits may be obtained.

When a solid sample is introduced into an atomizer it will not be in intimate contact with the graphite surfaces; therefore heating of the sample and reduction efficiency will be reduced. As a result of this, several workers[52-56] have found it advantageous in some cases to mix the sample with graphite powder prior to its introduction into the atomizer.

There are several additional factors to be considered from the analytical viewpoint which can limit the number of applications of solid sampling. (i) The concentration ranges of the trace elements which can be determined by direct atomization from the solid are limited by the weights of sample (0.1 to 10 mg) which can be introduced into the atomizer. (ii) The problem of heterogeneity is important in some cases, with sample weights being so small, although the problem can be minimized by finely grinding the samples.[57] (iii) Calibration is extremely difficult and usually requires standardization with similar, analysed, samples (see also Section 4.5.c.). (iv) With currently available spectrometers, solid sampling is not suited to multi-element analysis, because each determination requires an accurately weighed sample, compared to the solution procedure, where numerous determinations can be achieved from one weighed sample. (v) With the tantalum boat method of sample introduction, weighing before and after sample addition is frequently necessary if the sample is not transferred quantitatively to the furnace. (vi) With the cup method, calibration can be slightly more imprecise, since several different cups may be used, all of varying age and therefore all with slightly different performances. (vii) The precision of solids sampling is generally poorer than for liquid sampling.

In summary, with currently available instrumentation, solid sampling has several advantages but it is a procedure for special cases rather than the general method.

(b) Liquids

Liquid sampling, as with flame spectrometry, is the normal method of sample addition. However, with electrothermal atomization continuous sample introduction is the exception rather than the rule; instead, discrete samples are introduced with micropipettes capable of dispensing volumes of $0.5-100$ μl. Typical sample volumes are $1-10$ μl for cups, rods, and filaments and $5-50$ μl for furnaces. Micropipettes made completely of plastic and diposable plastic sampling-tips are the most suitable, since these considerably reduce any possibility of contamination. Contamination due to leaching from pipette tips can still occur, but this problem can be greatly reduced by rinsing the tips, before use, with hydrochloric or nitric acid, followed by distilled water.[57a,58] Not only can metal sampling-tips give rise to sample contamination but they could cause adsorption of certain elements from the sample solution, *e.g.* Reeves *et al.*[58a] have reported losses of Ag from aqueous solutions when using a microlitre syringe fitted with a stainless-steel needle.

Fixed-volume micropipettes, because of their ease of use and acceptable precision ($\pm 0.5\%$) and accuracy ($\pm 1\%$), are unlikely to be superseded, for routine analytical applications, by other more sophisticated techniques. However, attempts have been made to improve the precision and accuracy of sample addition in widely differing ways. Maessen *et al.*[59] have concentrated on the problem of reproducing the position of the sample solution on a graphite rod, using a specially designed pipette holder and optical alignment. Because of the temperature gradient existing along the atomizer during atomization of the sample, the positioning of the sample is of paramount importance if the highest accuracy and precision are required. Layman and Hieftje,[60] on the other hand, have developed an electronic system for accurately measuring the volume of sample added to a tungsten- or platinum-wire filament. Their measuring system depends on the addition of the sample to a warm filament, producing a rapid drop in filament temperature that results in a drop in the filament resistance, which gives rise to a drop in voltage. As the liquid sample evaporates the filament remains at this lower temperature until the sample has completely evaporated, at which time the filament returns to its high temperature, producing an increase in the voltage across the filament. The time between these two voltage changes, which can be accurately measured, is directly proportional to the sample volume over the range $0.5-5$ μl.

Of far greater practical importance to the analyst is the work of Pickford and Rossi,[61] who have produced an automated sampling system capable of repeated sampling from a flowing stream of water. The automatic sampling and injection unit of their apparatus is shown in Figure 4-1; the central programming system which controls this unit, together with the graphite furnace programmer and read-out, is not included. When the programme is started the two cams, which are rigidly attached to each other, begin to rotate. The lower, larger cam pushes the quartz capillary forward into the injection port of the graphite tube; when the tip of this capillary has reached the centre of the graphite tube the upper, smaller cam activates the sampling valve. Movement of the central slider by 1 cm transfers 100 μl from the sample loop to the argon stream and hence to the graphite tube. Low sample (20 ml min^{-1}) and argon (5 ml min^{-1}) flow-rates are used to prevent leakage when the valve is in the intermediate position during sampling and to avoid a tendency for the sample aliquots to break up in the graphite tube. After addition of the sample the valve returns to its start position and the programmer initiates the drying stage on the furnace control unit. At this point it is possible to proceed to the pyrolysis and atomization stages or to return to the beginning of the cycle and add further aliquots (up to four) of the sample in order to increase the sensitivity of the determination. The apparatus, with modification, could easily be linked to a conventional automatic sample changer and so increase its areas of application considerably.

The question of contamination and loss of trace elements from solution has been mentioned above in connection with the use of micropipettes. This question is of considerable importance when one remembers the high sensitivity of electrothermal atomization; it is an area which again can only be outlined briefly here. The work of two authors serves to illustrate the problems. Robertson has investigated the possibility of both contamination and loss of trace elements during the analysis of sea-water. Using neutron activation analysis, he determined the concentrations of Zn, Fe, Sb, Co, Cr, Sc, Ag, Cu, and Hf in a wide range of laboratory materials (*e.g.* glasses, plastics, rubber, and paper tissues), distilled water, acids, organic solvents, and chemical reagents.[62] Zinc (up to 2%) and Fe (up to 0.04%), as might be expected, proved to be the greatest contaminants. He also investigated the rate of loss of Sc, Fe, Zn, Co, Sr, Rb, Ag, In, Sb, Cs, and U from sea-water stored in Pyrex and polythene bottles.[63] Without acidification, losses of In, Sc, Fe, Ag, U, and Co were observed with both containers after only a few days; on acidification with hydrochloric acid, to pH 1.5, the losses were eliminated in all cases except for Sc. Struempler[64] has carried out a similar investigation into adsorption of Ag, Pb, Cd, Zn, and Ni, using borosilicate glass, polyethylene, and poly-

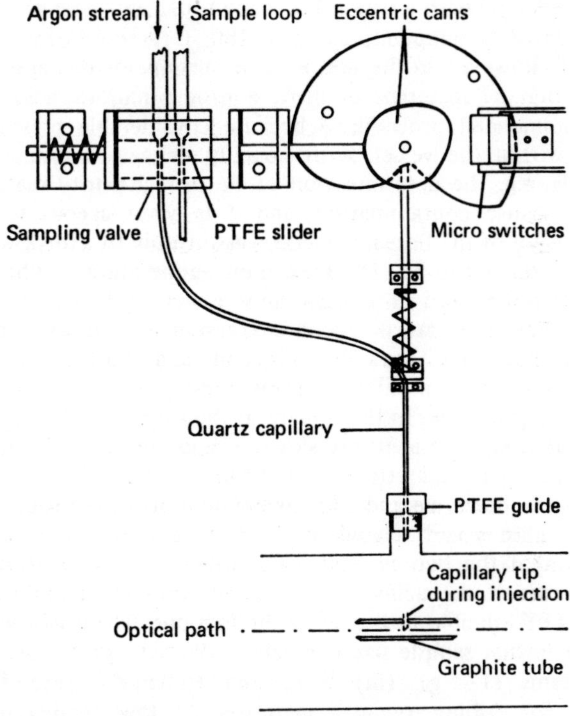

Figure 4-1   An automatic sampler for a graphite furnace atomizer (Reproduced from *Analyst,* 1972, **97**, 647).

propylene containers, in connection with the determination of these elements in rain, snow, and hail. He found that acidification of the samples to pH 2 with nitric acid in either borosilicate or polyethylene containers was the most acceptable, with polyethylene preferred because of its better handling characteristics.

Methods of solution produce additional problems when the samples are to be analysed by electrothermal atomization. Greater attention must be given to the quality and quantity of chemical reagents used, otherwise unacceptably high reagent blanks will be obtained. Three techniques are worthy of mention in this respect. (i) The PTFE pressure decomposition vessel of Bernas[65] facilitates solution procedures, enabling both inorganic and organic matrices to be dissolved, using the minimum quantities of acids. The vessel consists of a 25 ml PTFE crucible with lid contained in a stainless-steel shell with a removable screw top. The whole vessel is capable of being heated to temperatures up to 180 °C. A word of warning should be sounded, however, in the use of some versions of this apparatus when dissolving organic materials in nitric and/or perchloric acid (even using some recommended procedures) because considerable pressures can be generated within the vessel, with possibly dangerous results. For some applications, *e.g.* the determination of Fe in high-purity materials, these vessels can cause contamination, and it is advantageous to use a less versatile version of the Bernas vessel consisting only of a thick-walled PTFE crucible and screw-top lid. The maximum temperature at which this type of vessel can be used is considerably lower (120 °C) but it has the advantage that there are no metallic components. (ii) The vapour-phase acid-attack technique used by Mitchell and Nash[66] is particularly advantageous in that it allows cheaper, less pure, acids to be used for sample decompositions. In this apparatus the solvent acids and samples are kept separated and the acids are slowly evaporated, so that the generated vapours pass over, and attack, the sample. A simplified version of this apparatus has been designed by Woolley[67] for the decomposition of high-purity glasses and related materials. His apparatus is essentially a combination of the two systems, consisting of a vessel similar to that of Bernas, in which the acids are contained, with the sample placed in a separate PTFE crucible sitting inside the first one. The disadvantage of this equipment is that sample size is much smaller (0.1 g) compared with the other systems (1–3 g). (iii) Gleit and Holland[68] have developed a technique for ashing organic materials at low temperatures in an atmosphere of electronically excited oxygen. This technique removes the need to carry out dry ashing in a muffle, with the possible loss of volatile elements, and the alternative of using a wet oxidation procedure, with the resultant possibility of high reagent blanks.

The analysis of suspensions and emulsions is a promising method of sampling, combining the advantages of both solid and liquid sampling, but which has received surprisingly little attention. Providing a suitable medium can be found to disperse the finely ground sample, the method requires only one sample weighing, allows further dilutions to be made, and enables several determinations to be carried out on each suspension. Coupled with this, the method could avoid lengthy solution procedures and high blank values due to reagents. The main problems of the technique are possible sample contamination during grinding of the sample and the difficulty of obtaining a representative aliquot from the suspension to transfer to the atomizer for analysis.

(c) Gases

The direct determination of trace elements in gaseous samples has received very little attention; caused mainly by the necessity of operating the atomizers in an inert atmosphere. The exception has been the determination of As and Sb, after generation of their hydrides using sodium borohydride, because the liberated hydrides can be fed easily into the stream of inert gas flowing to the atomizer.

Analysis of trace elements in air particulates is, however, an area of considerable interest, using the cup and furnace designs; this topic has been included in this section, although strictly speaking it should come into the field of solid sampling. Two basic techniques of sampling are currently used. (*a*) Air samples are passed through filter papers with known pore sizes, which are then either wet oxidized and dissolved prior to analysis or are introduced directly into the atomizer. (*b*) Using the cup design, the graphite cup itself is used as the filter, and this is then inserted directly into the atomizer work head. In principle, the second technique is extremely attractive but it suffers from several of the disadvantages of direct solid sampling, *i.e.* single element determinations, limited analytical range on each sample, and the problem of non-representative sampling due to the small volumes of air required for analysis (0.1–10 l).

(d) Radioactive Samples

Since the capability of electrothermal atomizers to handle radioactive samples does not fit specifically into any one of the three categories above, special mention will be made of it here. The technique has the advantages, over flame atomic absorption, that the entire apparatus can easily be confined to a special enclosed work area, with remote-controlled sample handling and data feed-out systems if required.[68a] Exhaust fumes from the atomizer, containing the radioactive material, can relatively easily be monitored and controlled.

4.2 Separation Procedures

Separation procedures, to obtain lower detection limits and remove matrix interferences, which have been described in the literature for other analytical techniques can also be applied to methods which involve a final determination by electrothermal atomization. Stricter control is required, however, on reagent purity and analytical conditions due to the high sensitivity of this technique: see Section 4.1. Some research work has been carried out which was designed to study factors specifically affecting the use of organic chelating agents and organic solvents with electrothermal atomizers.

Table 4.1   Effect of drying parameters on the detection of 0.2 ng Co (50 μl; 0.004 μg ml^{-1}) in MIBK − APDC extracts. [69]

Time delay/ minutes	Drying temp./ °C	Drying time/ seconds	Peak height	
			1000 °C Pyrolysis	Atomization
0	130	60	46	29
0	100	100	50	32
0	50	120	61	38
0.5	130	60	68	43
2	130	60	73	47
4	130	60	76	47

Elements in organic solvents generally give identical results to aqueous standard solutions, although with some atomizer designs problems are encountered with larger volumes of organic solvents. This is caused by variations in the wetting properties between organic solvents and water, which cause variations in the extent to which the solvents spread on the atomizer surface. The increased spreading of organic solvents can cause variations in the rate of atomization due to the temperature gradient along the heated atomizers. This problem is minimized by using smaller sample volumes or improved atomizer designs. Dudas[69] has shown that variations in the time that elapses between sample introduction and the start of the atomizer heating cycle, together with variations in drying temperature, can seriously alter the analytical response obtained (Table 4.1).

Studies on the usual organic compounds and solvents used for extraction procedures[70-72] indicate that most complexes formed decompose to stable inorganic compounds, at low pyrolysis temperatures, giving the same sensitivity as that obtained with aqueous standards.

Separation of trace elements onto ion-exchange resins followed by the direct analysis of the resin has not received much attention as an analytical procedure, despite its obvious advantages. Muzzarelli and Rocchetti have demonstrated this technique with the determination of vanadium in sea-water.[73] They passed 1 l of sea-water through 500 mg of chitosan (a natural chelating polymer) and then analysed 5 mg samples of the homogenized resin for vanadium. The calibration graphs obtained for vanadium in the presence of the resin and in aqueous solution were identical.

An area of research which has received more attention than would seem to be justified from its limited analytical applications is the use of the electrodeposition preconcentration technique. Metal wires,[74-76a] graphite rods,[77] and hanging-mercury-drop electrodes[78,79] have been used to preconcentrate trace metals from solution. The metal wires and rods were subsequently used as the atomizer while the mercury drop was transferred (with some difficulty) to the atomizer for analysis. The application where this technique may be of some practical use is in the analysis of natural waters; the analytical procedures would be rather lengthy, and require an extremely competent analyst to carry them out.

4.3 Selection of Instrumental Parameters

(a) Light Sources

Selection of light sources, wavelengths, and slit widths are not peculiar to electrothermal atomization, and have been discussed adequately in the flame spectrometry literature. Therefore, except where relevant, they are not discussed here.

It is important when setting up the spectrometer to maximize the analytical signal against the light emission from the heated atomizer (an essential factor for atomic emission measurements). This entails operating with as low a gain setting on the photomultiplier as possible. From this aspect the higher light output of electrodeless discharge tubes gives them an advantage over hollow-cathode lamps as light sources for electrothermal atomization,[79a] particularly for atomic fluorescence measurements. However, since most analytical chemists still use atomic absorption and the more readily available hollow-cathode lamp as light source, it is often a considerable advantage in terms of detection limit to run the hollow-cathode lamp at slightly higher than normal operating currents. An alternative to this is to reduce the light intensity, from the atomizer, which reaches the photomultiplier; the variation of light emission from a heated graphite atomizer as a function of wavelength is shown in Figure 4-2. This can be achieved in two ways, either (a) by reducing the atomization temperature or (b) by fitting light baffles along the optical path of the

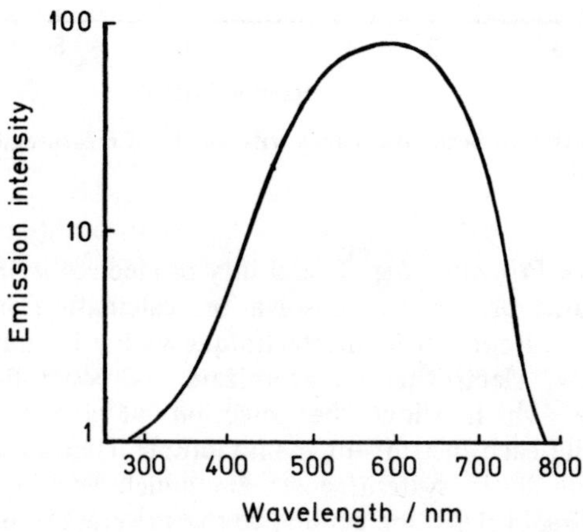

Figure 4-2   Variation of emission from a graphite furnace (2200 °C) with wavelength.

spectrometer, with an aperture slightly smaller than the normal aperture of the instrument.

(b) Atomizer Parameters

(i) Position. The positioning of cups and furnaces is only a matter of setting them up in the optical path of the spectrometer to allow the maximum light intensity to pass through. For rods and filaments the positioning is critical, because the atomic population decreases rapidly as the distance of measurement above the atomizer is increased; see Figure 4-3. Therefore for the highest analytical sensitivity the atomizer is normally adjusted to allow the light path to pass at grazing incidence over the rod or filament.

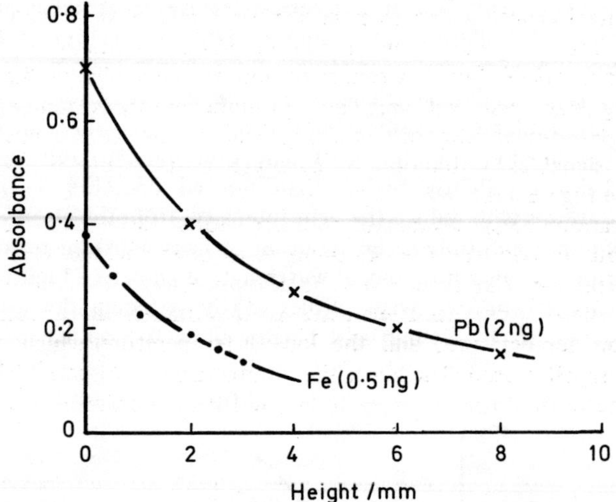

Figure 4-3   Variation of peak absorbance with height of measurement above a carbon filament

(ii) Temperature Programming. The ability of electrothermal atomizers to separate the flame processes of desolvation, calcination, and atomization imparts a degree of control to the technique which is unattainable with a flame. Generally, electrothermal atomizers are controlled by versatile power supplies which allow the selection of a range of operating conditions. While each instrument manufacturer has its own variation, all conform to one basic system; a system which enables three separate controlled heating cycles to be applied to the sample, corresponding to the so-called drying, pyrolysis, and atomization stages.

Drying stage: this is normally the least important stage and consists of a low-temperature heating cycle to evaporate any solvents: a temperature just below the boiling point of the solvent is ideal as this prevents sputtering of the sample. The drying time required, in seconds, can be calculated by a simple rule of thumb as being equal to 1½—2 times the sample volume used, measured in microlitres; *e.g.* a 20 μl sample would require a drying time of 30—40 s.

Pyrolysis stage: this is frequently the most important experimental parameter to be controlled, since by its judicious control it can considerably simplify an experimental determination; but by its injudicious selection it can lead to completely erroneous results. The aim of the pyrolysis stage is to heat the sample to the highest temperature possible without significant loss of the element being determined. In this way the sample matrix can frequently be broken down or vaporized prior to the atomization stage, thereby reducing problems caused by inter-element interferences and background light scattering or absorption. The pyrolysis time is dependent on the sample matrix (normally being longer for organic matrices) but as a generalization a time similar to that used in the drying stage will suffice. The optimum pyrolysis temperature is normally determined experimentally (see also page 32) by plotting the analytical signal as a function of pyrolysis temperature or instrument setting: see Figure 4-4.

Atomization stage: atomization temperatures vary from element to element, but in addition the optimum atomization temperature for any given element can vary from one type of sample to another. Normally the analytical signal for each element is determined as a function of atomization temperature, and the lowest temperature which gives the maximum signal is used. The lower the atomization temperature the longer the lifetime of the furnaces, cups, rods, and filaments. However, this is not

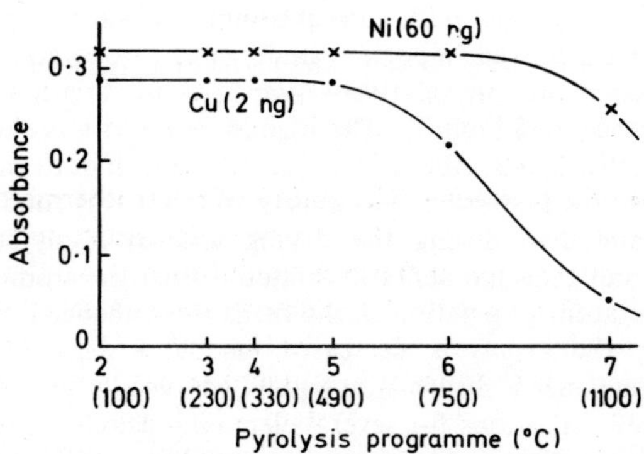

Figure 4-4   Variation of absorbance with pyrolysis temperature for a graphite furnace.

always the most suitable analytical temperature, *e.g.* if the matrix is relatively non-volatile. In these cases a higher than normal temperature is desirable to remove the element being determined from the sample matrix and to remove the matrix from the atomizer, to avoid an accumulation of non-volatile components. The atomization time should be the minimum required to vaporize the element completely, *i.e.* the analytical signal should return to zero.

The optimum pyrolysis and atomization temperatures for the elements are included in Chapter 5.

(c) Sheathing Gas

Electrothermal atomizers operate under conditions which minimize the rate at which the physical properties of the atomizers deteriorate; this normally entails operating in the absence of oxygen. Metal atomizers become brittle on oxidation while the surfaces of carbon atomizers rapidly become porous, due to loss of carbon as carbon monoxide or carbon dioxide.

A variety of stationary and flowing sheathing-gases have been used to achieve this result, together with other objectives.

(i) Inert Gases. This is the most common system, using oxygen-free argon or nitrogen. For absorption measurements, with carbon atomizers, nitrogen is normally completely adequate (and considerably cheaper than argon), although for metal atomizers argon is preferable, due to the possible formation of the metal nitride. Fluorescence measurements, with all types of atomizers, are better performed in an atmosphere of argon owing to the decreased quenching of the activated atoms compared to the diatomic gas nitrogen. In some cases (*e.g.* titanium, barium, molybdenum) reduced signals have been observed with nitrogen, due to the probable formation of nitrides.[80] With nitrogen there is the added consideration of the formation of cyanogen with carbon atomizers, but this problem does not appear to have proved important under normal operating conditions.

Signal measurements are relatively insensitive to variations in gas flow rate over the range $1-5$ l min^{-1}. The highest sensitivity is obtained under static gas conditions, *e.g.* the L'vov furnace, and modifications can be made to the gas flow systems of furnaces to simulate this condition.[24] The gas flow is maintained during the drying and pyrolysis stages of the determination and then momentarily halted during the atomization stage. In this way the atom population in the furnace is enhanced, since the rate of removal of the atoms is decreased (see also page 32). Table 4.2 illustrates the enhanced absorbance signals which can be obtained under stopped-gas-flow conditions for several elements, using a graphite furnace.

. Gas pressures, using a closed system, can be varied above or below atmospheric. Operating at increased pressure has two advantages: (*a*) the absorption profiles of atoms will be broadened in relation to the emission

Table 4.2 The effect of stopped gas flow on detection limits, using the HGA 70 graphite furnace.[24]

	Detection Limit/pg	
Element	Gas Flow	Stopped Flow
Ag	0.7	0.25
Bi	50	10
Co	20	4
Cu	2	1
Fe	6	3
Mn	1.5	1.0
Ni	20	10
Pb	15	6

lines from the light source, and (*b*) the diffusion rate of atoms from the atomizer will be reduced, giving rise to a longer residence time for the atoms.

(ii) Hydrogen Diffusion Flame. Amos *et al.*[81] showed that, by replacing the nitrogen or argon inert-gas shield around a carbon filament atomizer with an argon–hydrogen mixture, improved sensitivity could be obtained for many elements, together with a reduction in interference effects. The hydrogen in the gas mixture ignites spontaneously when the graphite rod is heated above a critical temperature and then burns as an argon–hydrogen–air-entrained diffusion flame. Table 4.3 illustrates the improvements obtained in the case of lead in terms of reduction in interferences and suppression of background absorption.

Table 4.3 The effect of an argon–hydrogen diffusion flame on the determination of lead with a graphite-rod atomizer.[81] Peak heights obtained with argon and argon–hydrogen flames are shown, for different concentrations of lead.

Interferent, 1000 μg ml^{-1}	Argon		Argon–hydrogen	
	1 μg ml^{-1}*	Blank	1 μg ml^{-1}*	Blank
–	100	0	100	0
H_3PO_4	32	19	103	0
NaCl	110	51	92	1
KCl	87	37	87	1½
$MgCl_2$	6	11	20	0
$CaCl_2$	0	46	92	1½

*Corrected for appropriate blank measurement

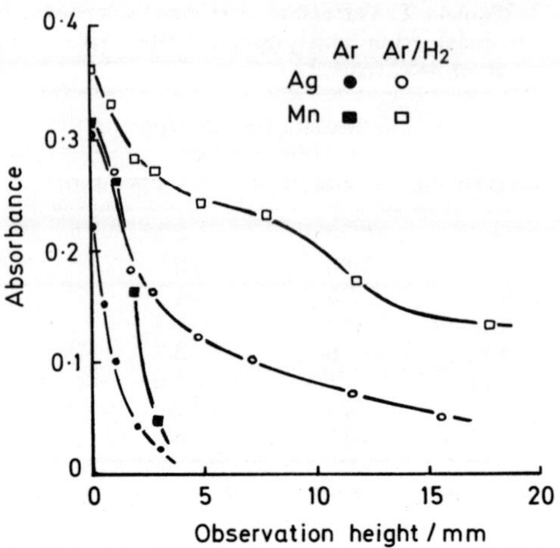

Figure 4-5   The effect of an argon–hydrogen diffusion flame on the decay of the atom population above a carbon rod atomizer.

Molnar and co-workers[82] have measured the temperature profile within the flame around the graphite rod and shown that the temperature does not exceed 500 °C above the rod. They concluded therefore that flame temperature does not contribute significantly to the atomization process.

The improvement in the carbon rod performance with an argon–hydrogen diffusion flame can be attributed to the highly reducing environment created, and the subsequent maintenance of the atomic population at much greater distances above the carbon rod than occurs with argon or nitrogen alone: see Figure 4-5. The reduced background signal has been attributed to a reduction in the presence of volatilized carbon due to hydrocarbon formation.[83]

(iii) Hydrocarbon–Inert Gas. Pyrolytically coated graphite atomizer components exhibit longer useful lifetimes and, frequently, improved analytical performances.[80,84,84a] This arises because the permeability of liquids and gases in the graphite is greatly reduced, thereby decreasing the possibility of liquid samples soaking into the atomizer, reducing the possibility of carbide formation, and reducing the rate of diffusion of gaseous atomic species through the walls of graphite furnaces. Graphite tubes and rods, however, can very soon lose this pyrolytic coating, together with its advantageous properties. These coatings could be replaced, using the apparatus of Siemer, Woodriff, and Watne[32] (see Chapter 2), but this is time-consuming and inconvenient. A much simpler procedure is to coat the atomizer pyrolytically *in situ,* either during a determination or between determinations. This technique uses the thermal

Table 4.4 The effect of methane additions to the purge gas on sensitivities obtained with a graphite rod.[80]

Element	Sensitivity/pg for 1% absorption	
	Argon	Argon–methane
Al	300	115
Ba	175	140
Be	3.5	2
Mo	25	25
Si	500	300[a]
Sn	650	250
Ti[b]	375	150
V	100	50

(*a*) Methane used between determinations.
(*b*) Nitrogen used as purge gas.

decomposition of a hydrocarbon at high temperature, which forms free carbon and hydrogen together with other decomposition products. The free carbon is deposited on the graphite atomizer to give a hard, grey, hydrophobic coating. Morrow and McElhaney[84] first used the method in 1973 by introducing an argon–methane (9+1) mixture (56 ml min^{-1}) into the normal argon purge gas (1.7 l min^{-1}) of a graphite furnace.

Thompson *et al.*[80] showed that methane, propane, ethylene, and acetylene additions of about 10 ml min^{-1} to the argon purge gas (2.5 l min^{-1}) flowing round a graphite rod all produced pyrolytic graphite coatings and improved the sensitivity for the determination of many elements (Table 4.4). The improvements were attributed to a reduction in carbide formation on the graphite rod and to the possible reduction in the free oxygen concentration around the rod due to the methane combustion. It is also possible that any free hydrogen formed would contribute to the increased sensitivity in the same way as that obtained with the hydrogen diffusion flame. Rod lifetime was also improved by a factor of 5 using this technique.

When a pyrolytic graphite coating is applied during an analytical determination, it is important to have a limited flow of hydrocarbon gas. If the gas flow is excessive, the high concentration of particulate carbon formed during the atomization stage can cause light scattering.

(iv) Oxygen–Inert Gas. Oxygen is normally excluded from the purge gas to avoid atomizer oxidation, particularly during the atomization stage. However, some authors[72,85,86] have suggested that the addition of oxygen during the pyrolysis stage of a determination could aid the oxidation and removal of organic matrices and therefore reduce the magnitude of background light scattering and absorption during the

atomization stage. This would be particularly advantageous in the determination of volatile elements, where high pyrolysis temperatures must be avoided.

Kundu and Prevot[86] showed that ashing at a temperature of only 490 °C for 90 seconds in the presence of 0.9 l min^{-1} nitrogen—oxygen (1+2) purge gas in a graphite furnace was sufficient to oxidize lipid samples containing compounds with chain lengths up to C_{18} completely. Without the addition of oxygen, pyrolysis temperatures in excess of 750 °C were required.

4.4 Background Correction

Problems arising from light scattering and molecular absorption are more frequent with electrothermal atomizers than they are with flames. Light scattering is caused by the sample matrix condensing, after vaporization, to form a smoke or mist; an event which occurs when a large quantity of vaporized material reaches the cooler region just above a rod or at the open ends of a furnace. Molecular absorption is caused through the broad absorption bands of molecular species present in the atomizer, particularly alkali-metal and alkaline-earth halides: see Figure 4-6 and Table 4.5.[20,87—89] Superimposed on these broad band absorption peaks can be sharp absorption lines due to the electronic spectra of molecules. It is important therefore to be aware of the advantages and disadvantages of the various methods of background correction in relation to the requirements of the transient signals obtained with electrothermal atomizers.

The three normal methods of background correction are well known from flame atomic absorption, *i.e.* (i) the deuterium arc of Koirtyohann and Pickett,[99] (ii) the deuterium or hydrogen hollow-cathode lamp, and (iii) the adjacent non-resonance line.

A fourth method of background correction currently receiving atten-

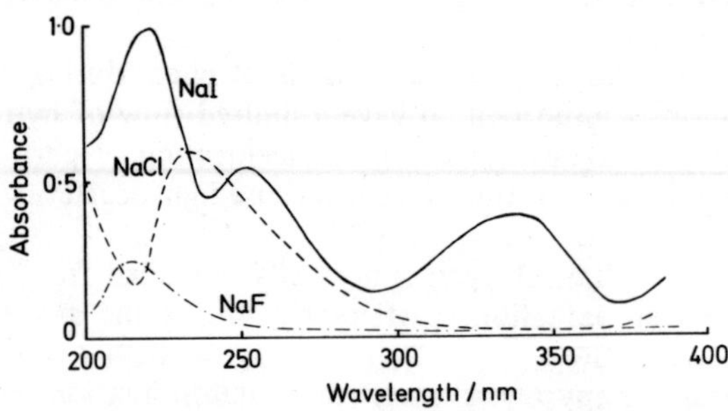

Figure 4-6   The molecular spectra of NaF, NaCl, and NaI (5 μg) in a carbon cup atomizer.

Table 4.5 Peak wavelengths, λ/nm, and peak absorbances, A,[a] in the molecular absorption spectra for sodium and potassium halides.

Compound	λ	A	λ	A	λ	A	λ	A
NaF	215	0.2	–	–	–	–	–	–
NaCl	<190	0.9[b]	235	0.8	–	–	–	–
KCl	195	1.0	250	0.7	–	–	–	–
NaBr	195	1.0	250	0.4	290	0.2	–	–
KBr	210	1.0	250	0.4	280	0.4	–	–
NaI	190	1.0	220	1.0	260	0.6	325	0.4
KI	195	1.0	240	1.0	260	0.6	320	0.4

(*a*) These values represent the absorbance reading obtained for 5 μg of each compound; the values given are only approximate, due to the non-reproducibility of the signals and to the variations observed with different types of electrothermal atomizer.
(*b*) Value obtained at 190 nm.

tion uses the Zeeman effect. This technique employs a magnetic field to split the spectral lines from the light source to give non-absorbable lines outside the atomic absorption profile. Dawson *et al.*[90a] have employed this technique with specific interest to electrothermal atomization. They constructed a vertical graphite rod atomizer housed between the pole pieces of an electromagnet. Zeeman splitting and polarization of the absorption lines in the atomic vapour itself could then be produced.

Simultaneous background correction is preferable to a sequential correction procedure because the background signals in electrothermal atomizers are normally less reproducible than with flames, and therefore vary from one determination to the next. Sequential correction procedures are often adequate, however; particularly if it is only required to show the absence of a background signal or if the background signal is small compared to the atomic signal.

The four techniques can be used to give simultaneous background correction, provided that the required spectrometer conditions are available. The deuterium arc is the most widely accepted of the four methods since it is readily incorporated into double-beam spectrometers and has a much higher light intensity over its useful range (200–360 nm) than the hydrogen or deuterium hollow-cathode lamps. The main disadvantages are that the geometries of the deuterium arc and analytical hollow-cathode lamp and the optical paths followed by the light from the two lamps are different. This makes it virtually impossible to superimpose the two light beams completely as they pass through or above the atomizer. The accurate correction of high background signals from a heterogeneously distributed smoke is therefore extremely difficult.

The hydrogen or deuterium hollow-cathode lamp system overcomes the problem of exactly matching the light beam with that from the analytical hollow-cathode lamp. Unfortunately, the light output from these lamps is low, which entails working with a higher photomultiplier gain setting, thereby increasing the noise signal. Also, superimposed on the continuum signal is the spectrum of the cathode material (*e.g.* nickel), which makes background correction for this element less reliable.

The adjacent non-resonance line technique suffers from none of the above problems; instead it has other, often more serious, problems. It is necessary to find a non-resonance line of adequate intensity within a few nanometres of that of the resonance absorbance line, and for simultaneous background correction it requires a dual-channel spectrometer. The most important argument against this method, however, is that no matter how close the non-resonance line, it cannot accurately measure the background signal at the wavelength of the resonance absorbance line. This is particularly true if the resonance line is situated on the side of a molecular absorption band. This method is particularly useful for background correction at wavelengths greater than 350 nm. Because the two-wavelength system is so universally applicable (*i.e.* it can be used with single-, double-, and dual-beam instruments and covers most elements), non-resonance lines suitable for background correction have been included with the experimental conditions for the determination of the elements in Chapter 5.

It must be remembered that all background-correction systems will correct for light scattering and broad-band molecular absorption but none will correct for background signals caused by molecular electronic vibrational spectra. The only method of correcting for this is by a sequential system using a blank corresponding to the sample matrix, and which is known to be free from the element being determined.

Background correction in atomic fluorescence measurements could be carried out by procedures corresponding to those described for atomic absorption. However, while balancing the light intensities from the resonance and non-resonance wavelength sources is relatively easy in flame fluorescence measurement,[91] it is more difficult with electrothermal atomizers.

Atomic emission measurements suffer from considerable difficulties caused by the intense background blackbody radiation from the hot graphite furnace. The technique of repetitive scanning or wavelength modulation has been used effectively to eliminate this problem of continuum radiation.[22a] This technique enables higher furnace temperatures to be employed, with the subsequent improvement in detection limits. With current commercial electrothermal atomizers the incorporation of wavelength modulation is almost indispensible if the best results are required.

4.5 Calibration

Calibration principles for electrothermal atomization are no different from other analytical techniques. Hence, calibration conditions should match the sample matrix as closely as possible, or a standard-additions calibration should be used. It is worth pointing out, however, that the standard-additions procedure is strictly only correct if the element being determined is added in the same form as that in which it is present in the sample. There are a number of additional points regarding calibration which are of particular importance to electrothermal atomization.

(*a*) Differences in signal response between standards in aqueous and organic solutions are in general much less than they are with flame atomization. Because the solvent medium is removed prior to atomization the only differences normally occurring are caused by variations in the spreading characteristics of the solvents on the atomizer prior to drying.

(*b*) Calibration graphs are usually prepared by plotting signal response against increasing solution concentration. If, instead, the signal response is plotted against the quantity of element present then the calibration graph can be prepared in a number of ways: (i) fixed volume at varying solution concentrations, (ii) fixed solution concentration at varying volumes, and (iii) fixed solution concentration and multiple fixed-volume additions, with sample drying after each volume addition. The differences obtained by these methods for aluminium, using a graphite furnace, are shown in Figure 4-7.[92] Of the methods

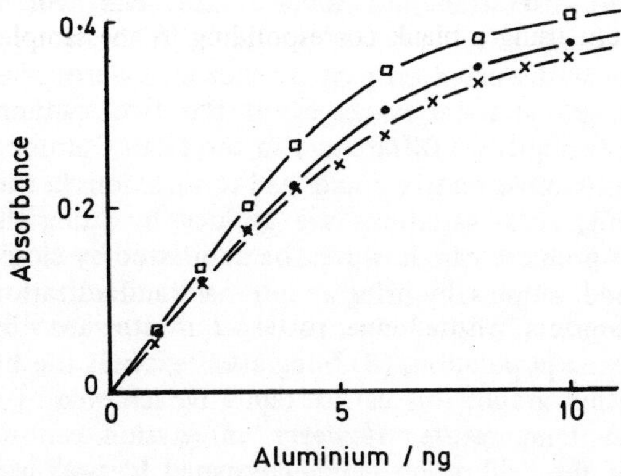

Figure 4-7   The effect of various sampling procedures on the calibration graph for aluminium, using a graphite furnace.
  □ — constant volume; variable concentration.
  ● — constant concentration; multiple aliquots of fixed volume.
  x — constant concentration; variable volume.

described, method (i) is superior to the others in terms of precision and accuracy, and is to be recommended.

(*c*) Solid samples present problems for calibration, and a universally acceptable method has not yet been proposed. Ideally, a series of standard samples is required, with increasing concentration levels for each element. A single standard material is not satisfactory since to prepare a calibration graph would require the addition of variable quantities of sample to the atomizer. This could result in variations of the heating efficiency of the sample during the atomization stage. Gries and Norval[93] proposed an ion-implantation technique for preparing primary metal standards. The method involved atomizing and ionizing the element required in an ion source of a heavy-ion accelerator and 'shooting' them into the metal matrix. The ionic charge transferred to the matrix was used to calculate the concentration of the trace element introduced. Unfortunately, this method has only limited applications, and would not be available to most analysts. A secondary solid standard proposed by the same authors[93] is considerably more practical, and deserving of further investigation. In this case a metal salt was introduced into molten urea, which was then cooled and ground to a fine powder. The authors claim that (i) urea can be obtained in a very pure state, (ii) many inorganic compounds can be introduced into molten urea by the proposed method, (iii) the standards are homogeneous, (iv) the matrix is readily pyrolysed in the atomizer, and (v) the resulting standards are stable at room temperature, with unlimited shelf-life if protected from excessive aerial humidity.

(*d*) Considerable discussion has taken place over whether peak or integrated signals give the better results for calibration purposes. This topic has been discussed on page 32, but it is worth considering here the advantages and disadvantages of the two methods from the practical viewpoint. (i) Differences in the rate of atomization due to variations in sample matrix could lead to variations in the peak height measurement; such variations are avoided by using the integrated signal. This problem can, however, be minimized by close matching of standards and samples, by using an internal standardization, or by using furnace atomizers, whose longer paths act as integrators by accumulating the atomic population. (ii) Integration extends the linear range of the calibration graph; this can certainly be achieved by reducing the atomization temperature. However, integration can only improve linearity of the calibration graph compared to peak-height measurements if the non-linearity is a function of poor instrumentation, *e.g.* a recording system with a slow response; it cannot remove non-linearity due to departure from Beer's law. Linearity of peak-height calibration graphs could be improved by using recorders that respond faster. In general, the rapid-heating rod and filament atomizers require recording systems with faster response than do furnaces if one is to avoid curvature of calibration graphs. (iii) Integration can give equal or greater sensitivity than peak-height measurements but the detection limits obtained by the two methods will be identical. (iv) Integration can lead to completely erroneous results if multiple peaks occur during the atomization stage (*e.g.* due to background signals). The individual signals cannot be easily isolated by the integrator system, whereas this is readily achieved by an analyst inspecting the recorded signals. So, even with electronic integration, a signal-recording system is still required to ensure a correct interpretation of the results. Both methods of calibration clearly have advantages, but on the balance of cost, convenience, and interpretation of results the recorded signal using peak-height measurements appears to be superior for routine operation.

4.6 Interferences

Interferences are the major problem in the use of electrothermal atomizers. Many authors have reported the occurrence of interference effects but very few of these authors have made a serious attempt at understanding or explaining the causes of these effects. Interference studies of the type which report the effect of A on B are useful for individual analytical problems, provided some attempt is made to overcome the effect. However, these studies have contributed very little to the general understanding of the problems, and considerably more work is required in this field.

It is worth pointing out that interferences are not necessarily absolute effects, since they can be dependent on the instrumentation used. For example; the presence of a matrix may merely slow down the release of the atomic species, resulting in a decrease of the peak-height signal; the integrated signal would, however, remain unchanged. If, on the other hand, the total number of atoms liberated is changed by the presence of a matrix then both the peak height and integrated signals will be changed.

In this section the various types of interferences are summarized under the two headings of 'physical' and 'chemical'.

(a) Physical Interferences

(i) Sample Introduction. Since there is a very steep temperature gradient along most atomizers, any variation in the sample size or position could give rise to a variation in signal response: see Section 4.1.b. These variations can be reduced in some instances by using signal integration: see Section 4.5.

(ii) Background Signals. Large concentrations of matrix vaporized during the atomization stage can cause scattering of the incident light beam. Molecular species vaporized during the atomization stage can cause

molecular absorption by their broad band and line spectra. All three effects give rise to apparent analytical signals which can be overcome in many cases by the use of background-correction techniques: see Section 4.5.

Alkali-metal halides generally give rise to the greatest molecular absorption, so their concentrations in the sample matrix should be kept to the minimum whenever possible.

Light emission from the heated atomizer can reach the photomultiplier and cause distortion of the baseline signal. This is particularly important when measurements are made in the visible region of the spectrum and when using low-intensity light sources. The occurrence of this effect can be identified by running the atomizer through the normal analysis cycle but without introducing the sample. If attenuation of the light beam does occur it can be overcome or reduced by the correct selection of instrument parameters: see Section 4.3.a.

(iii) Memory Effects. Incomplete atomization of an element causes an enhancement in subsequent analytical determinations; if the element is allowed to accumulate on the atomizer the errors become progressively greater. The greatest problem naturally occurs with those elements which form stable refractory oxides, *e.g.* V, Mo, W. Memory effects are diminished by using higher atomization temperatures or longer atomization times. It is sometimes advantageous to operate graphite atomizers at very high temperatures, so that the surface graphite just begins to vaporize, to ensure complete removal of the sample. Provided the surface coating of the graphite atomizer is replenished, by the pyrolysis of a hydrocarbon [see Section 4.3.c.(iii)], the performance of the atomizer is not affected seriously.

(b) Chemical Interferences
(i) Pyrolysis Losses. Losses of the analyte element can occur during the pyrolysis stage for two reasons: firstly, the element may be present in the sample in such a form that it is significantly volatile at the pyrolysis temperature employed, and secondly, the element may be converted into a volatile form by the sample matrix (either organic or inorganic). In either case it is usually advisable to check the recovery of an element from a particular matrix by using the method of standard additions. It is useful to study the effect of pyrolysis temperature on the analytical signal [see Section 4.3.b.(ii)], but again it is essential to carry out this study in the presence of the sample matrix, as the results obtained can be dependent on the other species present in the sample.

(ii) Anion/Cation Interferences. The presence of oxyanions in the sample matrix generally is preferred to halides because the loss of the analyte element is more likely to occur through vaporization of molecular halide species. While this generalization normally is true, it can be misleading, and

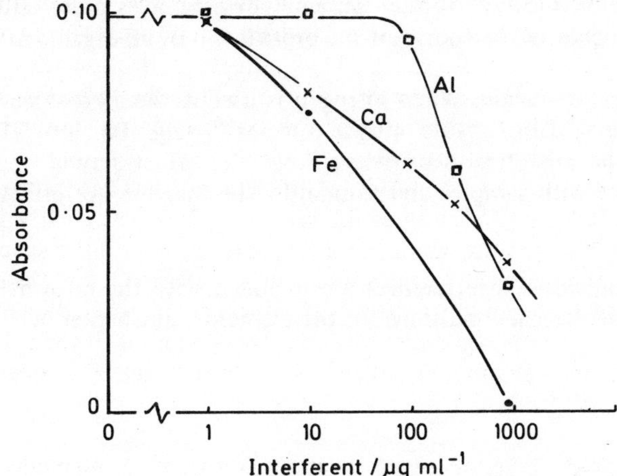

Figure 4-8   The effect of Al, Ca, and Fe (present as chlorides) on the absorbance
of 0.2 μg ml^{-1} of Pb (present as nitrate), using a graphite furnace.

should be confirmed for each analytical determination when possible.
Cation interferences are more complex, and no general theory has been
proposed yet to explain the many interferences observed: for example, see
Figure 4-8.

Anion/cation interferences can often be minimized by close matching
of standards to samples or by solvent-extraction procedures.

(iii) Condensation. When the atomized element leaves the hot surface of
an atomizer it is transported into a much cooler region where condensation of the atomic species can occur. This is enhanced by the presence of
large quantities of other elements that are atomized or vaporized at similar
temperatures to the element being determined since the analyte element
becomes occluded with the matrix as this condenses. This type of
interference is particularly prevalent with the filament type of atomizer.
However, the effect can be minimized by using limited-field viewing at
grazing incidence to the atomizer [see Section 4.3.b.(i)], or by using the
hydrogen-diffusion flame [see Section 4.3.c.(ii)].

(iv) Carbide Formation. The slow release of an element in the atomic form,
from graphite atomizers, is often ascribed to the formation of stable
carbide species; this is still a matter for debate. However, it is known that
for some elements the analytical signals obtained are very dependent on
the physical condition of the atomizer surface. To minimize any variation
in signal response it is essential to maintain the atomizer in a reproducible
condition, *e.g.* by the use of continuous pyrolytic coating. Metal atomizers

reduce the possibility of the element forming a carbide, although this could still occur in the course of the breakdown of an organic matrix.

(v) Nitride Formation. When nitrogen is used as the inert shield gas, some elements may form stable nitride species[80] (*e.g.* Ba, Mo, Ti), thereby reducing the analytical sensitivity. Since the effect should occur to the same extent with samples and standards, the interference effect is usually minimal.

Many individual interferences are included with the information given, under the analytical conditions for the elements, in Chapter 5.

5 Analytical Conditions for the Determination of the Elements by Atomic Absorption Spectrometry using Electrothermal Atomization

Various analytical parameters are given in this chapter for each of the elements which have been determined using electrothermal atomization. Values given for sensitivity, detection limit, maximum temperature of pyrolysis for a 30 second pyrolysis stage [T(pyrolysis)], and the optimum temperature of atomization [T(atomization)] normally refer to aqueous solutions. Within this chapter, all values quoted are of sensitivity/pg, detection limit/pg, T(pyrolysis)/°C, and T(atomization)/°C. The figures quoted are intended to be typical of those which should be obtainable with modern commercial graphite-furnace and tantalum-filament atomizers. It should be remembered that the values for sensitivities and detection limits are also dependent on the quality of the atomic absorption spectrometer. Spectrometer conditions used are normally those recommended by the instrument manufacturers for flame atomic absorption determinations, except that it is often advantageous to use slightly higher operating currents for hollow-cathode lamps and consequently a lower photomultiplier gain setting; see Section 4.3.a. Data on sensitivities refer to the quantity of an element required to give 1% absorption and data on detection limits refer to the quantity of an element required to produce an absorption signal equivalent to twice the noise of the baseline. Included for comparison are values for sensitivities and detection limits which are obtainable using the standard flame atomic absorption technique; these figures are based on a sample volume of 1 ml.

While simultaneous background correction using a continuum light source is the recommended technique to compensate for light scattering and molecular absorption effects [see Section 4.4], this is not always possible; for this reason suitable reference lines are included for each element to enable background correction by the two-line technique.

Interference studies with electrothermal atomizers have been very haphazard in their approach, and it would therefore be extremely difficult, and probably useless, to list all the results obtained for each element. This problem occurs because many interference effects which are apparent in one medium are absent in other media; reference is therefore given to the original papers for interference studies, and these should be consulted for the full details. As a generalization, where possible, analyses should be carried out in dilute oxyanion acid media; high concentrations of HNO_3, H_2SO_4, $HClO_4$, however, should be avoided as they lead to interference effects for most elements due to attack on the surface of the atomizer.

Non-specific interference effects [see Section 4.6] have not been included in this chapter. A large number of the interference studies reported with early prototype atomizers are not applicable to the commercially available atomizers and therefore have not been included here.

Aluminium

λ (resonance): 309.3 nm
λ (background): 307.0 nm
Stock solutions: aqueous — Al in the minimum quantity of HCl.
                organic — Al cyclohexanebutyrate and Al 2-ethyl-hexanoate in xylene or MIBK.

	Graphite furnace	*Tantalum filament*	*Flame*
Sensitivity	50 (*a*)	900 (*b*)	700 000
Detection limit	5	300	20 000
T(pyrolysis)	1400	1000	—
T(atomization)	2700	2000	—

(*a*) Enhanced sensitivity with CH_4. (*b*) Enhanced sensitivity with H_2.
Comments: Suppression by Fe, Na, and K.[94,94a] Enhancement by Ca.[80,95]

Antimony

λ(resonance): 217.6 nm (231.2 nm)
λ(background): 217.9 nm (Ni, 231.4 nm)
Stock solutions: aqueous — Sb in the minimum quantity of HNO_3–HCl or
                $2K(SbO)C_4H_4O_6,H_2O$ in water.
                organic —

	Graphite furnace	*Tantalum filament*	*Flame*
Sensitivity	20	800	1 000 000
Detection limit	5	200	500 000
T(pyrolysis)	1000	1000	—
T(atomization)	2000	2400	—

Comments: Detection limits are improved with the use of an electrodeless discharge lamp. Suppression by Al, As, Au, Bi, Co, Cr, Li, Mg, Ni, Pt and Se.[96–98] Enhancement by HCl, HNO_3, and H_2SO_4.[98a] Enhancement by Cr, Cu, and Ti.[99] Antimony has been determined by the generation of SbH_3 with $NaBH_4$ and passing the SbH_3 into a graphite furnace.[100,101]

Arsenic

λ(resonance): 193.7 nm (189.0, 197.3 nm)
λ(background): 192.0 nm
Stock solutions: aqueous — As_2O_3 in 0.02M-NaOH and KH_2AsO_4 in water.
                organic —

	Graphite furnace	Tantalum filament	Flame
Sensitivity	25 (a)	600	1 000 000
Detection limit	20 (a)	300	250 000
T(pyrolysis)	600	600	–
T(atomization)	2400	2400	–

(a) Using an electrodeless discharge lamp.

Comments: Detection limits are improved with the use of an electrodeless discharge lamp. Losses of As at elevated pyrolysis temperatures can be reduced by the addition of Ni.[102] Arsenic has been determined by the generation of AsH_3 with $NaBH_4$ and passing the AsH_3 into a graphite furnace;[100,101] sensitivity is improved by a factor of 10. A tantalum foil lining in a graphite tube has also been used to improve analytical sensitivity.[103] Suppression by PO_4, Bi, Co, Cr, K, Li, Sb, Se, Ti, and NH_4 halides.[96,97,98a,103a] Enhancement by Ca, Mg, Si, and SO_4[98a,103a] and by HCl, HNO_3, and H_2SO_4.[98a] Arsenic has been determined indirectly, after separation of the molybdo-arsenate, by determining Mo.[104]

Barium

λ(resonance): 553.6 nm
λ(background): Y, 557.6 nm and Ne, 540.0 nm
Stock solutions: aqueous – $BaCl_2$ in water.
         organic – Ba cyclohexanebutyrate in xylene or MIBK.

	Graphite furnace	Tantalum filament	Flame
Sensitivity	150 (a)	40 (b)	400 000
Detection limit	50	10	20 000
T(pyrolysis)	1600	1400	–
T(atomization)	3000	2000	–

(a) Enhanced sensitivity with CH_4. (b) Enhanced sensitivity with H_2.

Comments: Minimization of light, from the heated atomizer, reaching the photomultiplier is essential. Calcium additions (as the nitrate) of 10 000 μg ml^{-1} enhance sensitivity on a carbon rod by a factor of 5 and the use of Ar instead of N_2 improves sensitivity by a factor of 4.[105] A tantalum liner in a graphite furnace has been used to improve sensitivity.[106] Contamination can occur from the use of soft glass apparatus. Enhancement by Cs, K, and Na in HCl, by Al, Ca, K, and Na in HNO_3, and by H_3PO_4.[106a] Suppression by Ca, Fe, and La in HCl.[106a]

Beryllium

λ(resonance): 234.9 nm
λ(background): Sn, 235.4 nm
Stock solutions: aqueous – Be in the minimum quantity of HNO_3.
         organic – Be sulphonate in xylene.

	Graphite furnace	Tantalum filament	Flame
Sensitivity	2 (*a*)	7	20 000
Detection limit	0.5	2	1 000
T(pyrolysis)	1500	1500	—
T(atomization)	2700	2400	—

(*a*) Enhanced sensitivity with CH_4.

Comments: Enhancement by SO_4 and PO_4 and suppression by Ca.[107] Hydrogen has been found completely to suppress the signal for Be with a graphite furnace atomizer.[108]

Bismuth

λ(resonance): 223.1 nm (306.8 nm)
λ(background): Cd, 226.5 nm (Al, 307.0 nm)
Stock solutions: aqueous — Bi in the minimum quantity of HNO_3.
organic —

	Graphite furnace	Tantalum filament	Flame
Sensitivity	40	100	1 000 000
Detection limit	20	40	500 000
T(pyrolysis)	350	350	—
T(atomization)	1900	2200	—

Comments: Although the sensitivity is a quarter that of the 223.1 nm line, the 306.8 nm line intensity from the hollow-cathode lamp is about 20 times higher, and so it may provide a superior detection limit. Bismuth has been determined by the generation of BiH_3 with $NaBH_4$ and then passing the BiH_3 into a graphite furnace.[100,101]

Bromine

λ(resonance): 148.9 nm
λ(background): N, 149.3/149.5 nm
Stock solutions: aqueous — NH_4Br in water.
organic —

	Graphite furnace	Tantalum filament	Flame
Sensitivity	100	—	—
Detection limit	1000	—	—
T(pyrolysis)	700	—	—
T(atomization)	2500	—	—

Comments: Specialized equipment is required, consisting of a graphite tube furnace with LiF lenses as windows, a photoemission-scintillation detector, and a vacuum spectrometer.[109]

Cadmium

λ(resonance): 228.8 nm
λ(background): 226.5 nm
Stock solutions: aqueous — Cd in the minimum quantity of HNO_3.
organic — Cd cyclohexanebutyrate in xylene or MIBK.

	Graphite furnace	*Tantalum filament*	*Flame*
Sensitivity	1	7 (*a*)	30 000
Detection limit	0.1	2	1 000
T(pyrolysis)	350	350	—
T(atomization)	1900	700	—

(*a*) Enhanced sensitivity with H_2.

Comments: Detection limits for Cd are improved by the use of an electrodeless discharge lamp. Due to the rapid rate of atomization for Cd, a recording system with a fast response is required to avoid excessive curvature of the calibration graph. Sample solutions can be contaminated due to leaching of Cd from some types of micropipette plastic tips; these plastic tips should be pre-cleaned with dilute HNO_3. Cadmium determinations are prone to interference from background molecular absorption signals.

The presence of Cl in samples can cause the loss of Cd, as the volatile $CdCl_2$ species, during pyrolysis.[110-112] The loss of Cd can be reduced, however, by the addition of PO_4, SO_4, or F to the sample.[102,113] Many elements suppress the Cd response; Co, Cu, Fe, Li, Mg, Mn, Mo, Ni, Pd, Pt, Sr, and Zr.[111,112]

Higher pyrolysis temperatures, particularly important for organic matrices, can be used for Cd determinations by allowing oxygen to enter the atomizer during the pyrolysis stage.[114]

Calcium
λ(resonance): 422.7 nm
λ(background):
Stock solutions: aqueous — $CaCO_3$ in the minimum quantity of HNO_3.
           organic — Ca cyclohexanebutyrate or Ca 2-ethyl-
             hexanoate in xylene or MIBK.

	Graphite furnace	*Tantalum filament*	*Flame*
Sensitivity	4	3	50 000
Detection limit	20	1	1 000
T(pyrolysis)	1100	1100	—
T(atomization)	2600	2200	—

Comments: Minimization of light, from the heated atomizer, reaching the photomultiplier is essential.

Chromium
λ(resonance): 357.9 nm
λ(background): Ne, 352.0 nm
Stock solutions: aqueous — potassium chromate in water.
          organic — Cr cyclohexanebutyrate or tris(1-phenyl-
            1,3-butanediono)Cr in xylene or MIBK.

	Graphite furnace	Tantalum filament	Flame
Sensitivity	20	40 (a)	100 000
Detection limit	10	20	3 000
T(pyrolysis)	1200	1200	—
T(atomization)	2700	2400	—

(a) Enhanced sensitivity with H_2.

Comments: Suppression has been observed by Ca, Co, Fe, Mg, Ni, and Sr[115,116] and enhancement by Al,[117] KCl, Na_2SO_4, and also by $MgCl_2$.[118] Sodium, K, Ca, Cu, Zn, Fe, Mg, Mn, Cl, I, CO_3, HPO_4, and SO_4 have also been reported not to interfere at levels up to 5000-fold excess.[119] Enhanced sensitivity has been reported with CH_4 in the purge gas.[84]

Cobalt

λ(resonance): 240.7 nm
λ(background): 239.3 nm and Sn, 242.1 nm
Stock solutions: aqueous — Co in the minimum quantity of HNO_3.
                    organic — Co cyclohexanebutyrate in xylene or MIBK.

	Graphite furnace	Tantalum filament	Flame
Sensitivity	40	900 (a)	150 000
Detection limit	5	300	10 000
T(pyrolysis)	1100	1000	—
T(atomization)	2600	2000	—

(a) Enhanced sensitivity with H_2.

Comments: Suppression has been observed by NaCl,[120] Ca, Fe, Mg, and Mn.[121]

Copper

λ(resonance): 324.8 nm
λ(background): 323.1 nm
Stock solutions: aqueous — Cu in the minimum quantity of HNO_3.
                    organic — bis(1-phenyl-1,3-butanediono)Cu or Cu
                    cyclohexanebutyrate in xylene or MIBK.

	Graphite furnace	Tantalum filament	Flame
Sensitivity	30	20	100 000
Detection limit	2	10	2 000
T(pyrolysis)	800	800	—
T(atomization)	2600	1800	—

Comments: Suppression by Ag, Al, Ba, Ca, Cr, Fe, Li, Sr, SO_4, and PO_4.[94,122–124] Suppression by Ca, K, La, Mg, and Na chlorides[124a] also reported.

Erbium
λ(resonance): 400.8 nm
λ(background): 394.4 nm
Stock solutions: aqueous — Er in the minimum quantity of HNO_3.
            organic —

	Graphite furnace	Tantalum filament	Flame
Sensitivity	100 (*a*)	—	1 000 000
Detection limit	40	—	1 000 000
T(pyrolysis)	1500	—	—
T(atomization)	2700	—	—

(*a*) Using a Woodriff furnace.[12]

Europium
λ(resonance): 459.4 nm
λ(background):
Stock solutions: aqueous — Eu_2O_3 in the minimum quantity of HCl.
            organic —

	Graphite furnace	Tantalum filament	Flame
Sensitivity	40 000	70 (*a*)	1 000 000
Detection limit	40 000	30	1 000 000
T(pyrolysis)	1200	1000	—
T(atomization)	2700	2200	—

(*a*) Enhanced sensitivity with H_2.
Comments: Suppression by Al, Fe, and Zn; enhancement by Ca, Co, Cr, K, Mg, Mn, Na, and Ni.[125]

Gallium
λ(resonance): 287.4 nm
λ(background): Cd, 283.7 nm; Sn, 283.9 nm
Stock solutions: aqueous — Ga in the minimum quantity of HNO_3–HCl.
            organic —

	Graphite furnace	Tantalum filament	Flame
Sensitivity	400	300	2 000 000
Detection limit	200	100	70 000
T(pyrolysis)	700	700	—
T(atomization)	2500	1800	—

Comments: Enhancement by HNO_3.[102]

Germanium
λ(resonance): 265.2 nm
λ(background):
Stock solutions: aqueous — Ge in the minimum quantity of HF–HNO_3.
            organic —

	Graphite furnace	Tantalum filament	Flame
Sensitivity	100	—	2 000 000
Detection limit	300	—	1 000 000
T(pyrolysis)	800	—	—
T(atomization)	2700	—	—

Comments: Enhancement by HNO_3, H_2SO_4, and $HClO_4$.[102] Germanium has been determined by the generation of GeH_4 with $NaBH_4$ and passing the GeH_4 into a graphite furnace.[101]

Gold

λ(resonance): 242.8 nm (267.6 nm)
λ(background): Sn, 242.1 nm
Stock solutions: aqueous — Au in the minimum quantity of HNO_3–HCl.
                 organic —

	Graphite furnace	Tantalum filament	Flame
Sensitivity	20	100	300 000
Detection limit	10	20	100 000
T(pyrolysis)	500	500	—
T(atomization)	2400	2400	—

Comments: Although the sensitivity is half that of the 242.8 nm line, the 267.6 nm line-intensity from a hollow-cathode lamp is about 50 per cent higher, and so it may provide a superior detection limit.

Indium

λ(resonance): 303.9 nm
λ(background): Al, 307.0 nm
Stock solutions: aqueous — In in the minimum quantity of HNO_3.
                 organic —

	Graphite furnace	Tantalum filament	Flame
Sensitivity	40	300	1 000 000
Detection limit	20	100	50 000
T(pyrolysis)	700	700	—
T(atomization)	2000	1800	—

Iodine

λ(resonance): 183.0 nm (206.2 nm)
λ(background): 184.4 nm
Stock solutions: aqueous — NH_4I in water.
                 organic —

	Graphite furnace	Tantalum filament	Flame
Sensitivity	1000	—	12 000 000
Detection limit	4000	—	25 000 000
T(pyrolysis)	500	—	—
T(atomization)	2000	—	—

Comments: Argon-flushed optics required, with preferably a vacuum spectrometer.

Iridium
λ(resonance): 264.0 nm
λ(background):
Stock solutions: aqueous — Ir_2O_3 in the minimum quantity of H_2SO_4.
        organic —

	Graphite furnace	Tantalum filament	Flame
Sensitivity	500	—	10 000 000
Detection limit	1000	—	2 000 000
T(pyrolysis)	1200	—	—
T(atomization)	2700	—	—

Comments: Suppression by Au, Pd, Pt, Rh, and Ru.[125a]

Iron
λ(resonance): 248.3 nm (372.0 nm)
λ(background): Cu, 249.2 nm
Stock solutions: aqueous — Fe in the minimum quantity of HNO_3 or
                H_2SO_4.
        organic — tris(1-phenyl-1,3-butanediono)Fe or Fe
        cyclohexanebutyrate in xylene or MIBK.

	Graphite furnace	Tantalum filament	Flame
Sensitivity	25 (a)	200	100 000
Detection limit	5	100	10 000
T(pyrolysis)	1200	1200	—
T(atomization)	2500	2400	—

  (a) Enhanced sensitivity with CH_4.
Comments: Suppression by Ag, Au,[126] and Cr.[127]

Lead
λ(resonance): 283.3 nm (217.0 nm)
λ(background): 280.1 nm and Cd, 283.7 nm (220.4 nm)
Stock solutions: aqueous — Pb in the minimum quantity of HNO_3 or
                $Pb(NO_3)_2$ in water.
        organic — Pb cyclohexanebutyrate in xylene or MIBK.

	Graphite furnace	Tantalum filament	Flame
Sensitivity	20	30	500 000
Detection limit	2	10	20 000
T(pyrolysis)	600	600	—
T(atomization)	2100	1400	—

Comments: More work has been published on the determination of Pb
than for any other element; despite this, no clear or comprehensive study
has been made of the numerous interference effects observed. It is apparent,
however, that the most severe interferences arise from alkali-metal and
alkaline-earth chlorides.

Suppression has been reported by Ag, Al, Ba, Ca, Cd, Co, Cu, Fe, Hg, K,
Li, Mg, Mn, Na, Ni, Sr, and Zn.[117,118,122,123,128–130] One thousand-
fold excesses of Ba, Ca, Co, Cr, Cu, Fe, Mg, Na, Ni, and Zn, however,

have been reported not to interfere when present as their nitrates.[131]
Suppression has also been reported by HNO_3, H_2SO_4, and H_3PO_4,[94,132]
although H_3PO_4 can be used as an additive to samples of variable
composition to obtain reproducible results. Interference from SO_4 can be
reduced by the addition of 0.75% m/V La.[132a] Substantial enhancement
by edta has been reported.[133,134] Similarly, additions of ascorbic acid,
tartaric acid, and sucrose have produced increased sensitivity and
reductions in the interference effects caused by alkali-metal and alkaline-
earth halides.[134a]

Lithium

λ(resonance): 670.8 nm
λ(background):
Stock solutions: aqueous — Li_2CO_3 in the minimum quantity of HNO_3.
     organic — Li cyclohexanebutyrate in xylene or MIBK.

	Graphite furnace	*Tantalum filament*	*Flame*
Sensitivity	10	7	30 000
Detection limit	5	2	10 000
T(pyrolysis)	1000	1000	—
T(atomization)	2500	2000	—

Comments: Minimization of light, from the heated atomizer, reaching the
photomultiplier is essential. A detection limit of 4×10^{-12} g has been
obtained, with a graphite furnace, by atomic emission at 2700 °C.[22]

Magnesium

λ(resonance): 285.2 nm
λ(background): Cd, 283.7 nm and Sn, 283.9 nm
Stock solutions: aqueous — $MgCO_3$ in the minimum quantity of HNO_3.
     organic — Mg cyclohexanebutyrate in xylene or MIBK.

	Graphite furnace	*Tantalum filament*	*Flame*
Sensitivity	0.2	0.2	7 000
Detection limit	0.02	0.1	100
T(pyrolysis)	1000	600	—
T(atomization)	2200	1400	—

Comments: Difficulties in determining Mg by electrothermal atomization
frequently occur due to the problem of obtaining water of sufficient
purity.

Manganese

λ(resonance): 279.5 nm
λ(background): Pb, 280.1 nm, Cu, 282.4 nm
Stock solutions: aqueous — Mn in the minimum quantity of HNO_3.
     organic — Mn cyclohexanebutyrate in xylene or MIBK.

	Graphite furnace	Tantalum filament	Flame
Sensitivity	2	10	50 000
Detection limit	0.2	2	2 000
T(pyrolysis)	1100	900	–
T(atomization)	2600	1800	–

Comments: Suppression by Ba, Ca, Cu, La, Mg, Ni, Pb, and Sr when present as chlorides[124,124a,135] and by Cr.[127]

Mercury

λ(resonance): 253.7 nm

λ(background): Cu, 249.2 nm

Stock solutions: aqueous – Hg in the minimum quantity of HNO_3.

                    organic – Hg cyclohexanebutyrate in xylene or MIBK.

	Graphite furnace	Tantalum filament	Flame
Sensitivity	200	800	10 000 000
Detection limit	100	200	500 000
T(pyrolysis)	–	–	–
T(atomization)	850	900	–

Comments: Loss of Hg from samples during the pyrolysis stage is an extremely severe problem, even at normal sample drying temperatures of 100 °C. Losses of Hg during the pyrolysis and drying stages are reduced by the addition of $(NH_4)_2 S$.[102] Lech, Siemer, and Woodriff[136,137] have shown that gold-plating of a carbon atomizer also considerably reduces this problem. The Hg forms an amalgam with the gold, thereby reducing the vapour pressure of Hg. Atomization should be carried out at temperatures below the melting point of Au to maintain the Au layer. L'vov and Khartsyzov[138] have used a vacuum monochromator and an Ar-purged graphite furnace to determine Hg at 184.9 nm.

Molybdenum

λ(resonance): 313.3 nm

λ(background): 311.2 nm

Stock solutions: aqueous – $(NH_4)_6 Mo_7 O_{24}, 4H_2O$ in 10% HCl.

                  organic –

	Graphite furnace	Tantalum filament	Flame
Sensitivity	20	–	400 000
Detection limit	5	–	200 000
T(pyrolysis)	1900	–	–
T(atomization)	2700	–	–

Comments: Long atomization periods at high temperatures are essential to ensure complete removal of Mo from the atomizer.

Nickel

λ(resonance): 232.0 nm

λ(background): 231.4 nm

Stock solutions: aqueous – Ni in the minimum quantity of HNO_3.

                  organic – Ni cyclohexanebutyrate in xylene or MIBK.

	Graphite furnace	Tantalum filament	Flame
Sensitivity	100 (a)	400	100 000
Detection limit	20	200	10 000
T(pyrolysis)	1000	1000	—
T(atomization)	2700	2400	—

(a) Enhanced sensitivity with CH_4.
Comments: Suppression by Cr.[127]

Osmium

λ(resonance): 290.9 nm
λ(background):
Stock solutions: aqueous — OsO_4 in 2M-NaOH.
                 organic —

	Graphite furnace	Tantalum filament	Flame
Sensitivity	1000	—	1 000 000
Detection limit	2000	—	100 000
T(pyrolysis)	1700	—	—
T(atomization)	2500	—	—

Palladium

λ(resonance): 247.6 nm
λ(background): Cu, 249.2 nm
Stock solutions: aqueous — Pd in the minimum quantity of HNO_3–HCl.
                 organic —

	Graphite furnace	Tantalum filament	Flame
Sensitivity	20	2000	200 000
Detection limit	20	700	20 000
T(pyrolysis)	1400	1200	—
T(atomization)	2600	2400	—

Comments: Although the sensitivity is a fifth that of the 247.6 nm line, the 340.5 nm line intensity from the hollow-cathode lamp is about 20 times higher, and so it may provide a superior detection limit. Enhancement by Au, Ir, Pt, and Rh.[125a]

Phosphorus

λ(resonance): 177.5 nm (213.5/213.6 nm)
λ(background): — (Zn, 212.5 nm)
Stock solutions: aqueous — $(NH_4)_2HPO_4$ in water.
                 organic — triphenyl phosphate in xylene or MIBK.

	Graphite furnace	Tantalum filament	Flame
Sensitivity	4	—	—
Detection limit	4	—	—
T(pyrolysis)	500	—	—
T(atomization)	1600	—	—

Comments: Argon-flushed optics required, with (preferably) a vacuum spectrometer. A detection limit of 200 pg has been obtained at 213.5/ 213.6 nm. A detection limit of 4 ng has been obtained by emission from the HPO species at 526 nm, using a carbon rod atomizer with a hydrogen diffusion flame.[139] Phosphorus has been determined indirectly, after separation of the molybdophosphate, by determining Mo.[104]

Platinum

λ(resonance): 266.0 nm
λ(background):
Stock solutions: aqueous — Pt in the minimum quantity of HNO_3–HCl.
            organic —

	Graphite furnace	Tantalum filament	Flame
Sensitivity	500	—	2 000 000
Detection limit	200	—	100 000
T(pyrolysis)	1400	—	—
T(atomization)	2700	—	—

Comments: Suppression by Au, Ir, Pd, Rh, and Ru.[125a]

Plutonium

λ(resonance): 420 nm
λ(background):
Stock solutions: aqueous —
            organic —

	Graphite furnace	Tantalum filament	Flame
Sensitivity	—	—	—
Detection limit	100 000 (*a*)	—	—
T(pyrolysis)	1200	—	—
T(atomization)	2700	—	—

  (*a*) See ref. 140

Potassium

λ(resonance): 766.5 nm
λ(background):
Stock solutions: aqueous — K_2CO_3 in the minimum quantity of HNO_3.
            organic — K cyclohexanebutyrate in xylene or MIBK.

	Graphite furnace	Tantalum filament	Flame
Sensitivity	5	3	20 000
Detection limit	1	1	5 000
T(pyrolysis)	1000	1000	—
T(atomization)	2200	1800	—

Comments: Minimization of light, from the heated atomizer, reaching the photomultiplier is essential. A detection limit of 0.08 pg has been obtained, with a graphite furnace, by atomic emission at 2700 °C.[22] Losses of K can occur in the presence of Cl during pyrolysis.

Rhenium
λ(resonance): 346.0 nm
λ(background): Ne, 352.0 nm
Stock solutions: aqueous — $KReO_4$ in water.
                    organic —

	Graphite furnace	Tantalum filament	Flame
Sensitivity	1000	3000	12 000 000
Detection limit	1000	3000	2 000 000
T(pyrolysis)	1200	1000	—
T(atomization)	2700	2400	—

Rhodium
λ(resonance): 343.5 nm
λ(background): 350.7 nm and Ne, 352.0 nm
Stock solutions: aqueous — $Rh_2(SO_4)_3,4H_2O$ in water.
                    organic —

	Graphite furnace	Tantalum filament	Flame
Sensitivity	50	—	350 000
Detection limit	50	—	30 000
T(pyrolysis)	1200	—	—
T(atomization)	2500	—	—

Comments: Suppression by Ir, Pt, and Ru.[125a]

Rubidium
λ(resonance): 780.0 nm
λ(background):
Stock solutions: aqueous — Rb_2CO_3 in the minimum quantity of HNO_3.
                    organic —

	Graphite furnace	Tantalum filament	Flame
Sensitivity	20	20	50 000
Detection limit	10	10	5 000
T(pyrolysis)	1000	1000	—
T(atomization)	2400	2200	—

Comments: Minimization of light, from the heated atomizer, reaching the photomultiplier is essential.

Ruthenium
λ(resonance): 349.9 nm
λ(background): Ne, 352.0 nm
Stock solutions: aqueous — $RuCl_3$ in the minimum quantity of
                    HNO_3–HCl.
                    organic —

	Graphite furnace	Tantalum filament	Flame
Sensitivity	100	–	1 000 000
Detection limit	150	–	100 000
T(pyrolysis)	1700	–	–
T(atomization)	2500	–	–

Comments: Suppression by Au, Ir, Pd, Pt, and Rh.[125a]

Selenium

λ(resonance): 196.0 nm
λ(background): 198.1 nm
Stock solutions: aqueous – Se in the minimum quantity of HNO_3.
                 organic –

	Graphite furnace	Tantalum filament	Flame
Sensitivity	200	700 (a)	1 000 000
Detection limit	100	700	250 000
T(pyrolysis)	700	700	–
T(atomization)	2500	2000	–

(a) Enhanced sensitivity with H_2.

Comments: Detection limits are improved with the use of an electrodeless discharge lamp. Suppression by HCl, HNO_3, and H_2SO_4.[98a] Suppression by Be, Ca, NO_3, Na, SO_4, Si, and Sn.[103a,141] Enhancement by Ca, Cr, Cu, Fe, Ge, Hg, K, Mo, Na, Ni, Sb, Sr, Te, Ti, V, W, Zn, and Zr.[103a,141] The suppressive effects of 25 elements in 1% HNO_3 (Ag, Al, As, B, Ba, Be, Cd, Co, Cr, Cu, Fe, Li, Mo, Mn, Na, P, Pb, Sb, Si, Sn, Sr, Ti, Tl, V, Zn) on the Se signal was reduced by the addition of up to 1% Ni.[142] Losses of Se at elevated pyrolysis temperatures can be reduced by the addition of Mo,[141] Cu,[102] and Ni.[102,142–144] Improved sensitivity has also been achieved by lining a graphite atomizer with tantalum foil.[103] Selenium has been determined by the generation of SeH_2 with $NaBH_4$ and passing the SeH_2 into a graphite furnace.[100,101]

Silicon

λ(resonance): 251.6 nm
λ(background): Cu, 249.2 nm
Stock solutions: aqueous – $Na_2SiO_3,9H_2O$ in water.
                 organic – octaphenyltetrasiloxane in xylene or MIBK.

	Graphite furnace	Tantalum filament	Flame
Sensitivity	50	40 000 (a)	1 000 000
Detection limit	20	10 000	80 000
T(pyrolysis)	1200	1 200	–
T(atomization)	2700	2 400	–

(a) Enhanced sensitivity with H_2.

Comments: Calcium additions (as the nitrate) of 1000 $\mu g\ ml^{-1}$ enhance the sensitivity on a graphite rod by a factor of 3.[80]

Silver

λ(resonance): 328.1 nm (338.3 nm)

λ(background): Ne, 332.4 and Sn, 326.2 nm

Stock solutions: aqueous — $AgNO_3$ in water.

                organic — Ag cyclohexanebutyrate and Ag 2-ethyl-hexanoate in xylene or MIBK.

	Graphite furnace	Tantalum filament	Flame
Sensitivity	5	20	50 000
Detection limit	0.1	4	5 000
T(pyrolysis)	450	350	—
T(atomization)	2500	1200	—

Comments: Solutions containing low Ag concentrations are unstable, Ag is also adsorbed onto container walls. Solutions can be stabilized by 0.01% HNO_3.

Sodium

λ(resonance): 589.0 nm

λ(background):

Stock solutions: aqueous — Na_2CO_3 in the minimum quantity of HNO_3.

                organic — Na cyclohexanebutyrate in xylene or MIBK.

	Graphite furnace	Tantalum filament	Flame
Sensitivity	1	3	20 000
Detection limit	0.2	1	2 000
T(pyrolysis)	700	700	—
T(atomization)	2000	1600	—

Comments: Difficulties in determining Na by electrothermal atomization occur due to the problem of sample contamination and obtaining water of sufficient purity. Minimization of light, from the heated atomizer, reaching the photomultiplier is essential. A detection limit of 0.1 pg has been obtained, with a graphite furnace, by atomic emission at 2700 °C.[22] Loss of Na in the presence of Cl can occur during pyrolysis.

Strontium

λ(resonance): 460.7 nm

λ(background):

Stock solutions: aqueous — $SrCO_3$ in the minimum quantity of HNO_3.

                organic — Sr cyclohexanebutyrate in xylene or MIBK.

	Graphite furnace	Tantalum filament	Flame
Sensitivity	20	40	100 000
Detection limit	5	10	10 000
T(pyrolysis)	1500	1200	—
T(atomization)	2700	2200	—

Comments: Minimization of light, from the heated atomizer, reaching the photomultiplier is essential.

Sulphur

λ(resonance): 180.7 nm
λ(background):
Stock solutions: aqueous $-$ $(NH_4)_2SO_4$ in water.
       organic $-$

	Graphite furnace	Tantalum filament	Flame
Sensitivity	100	$-$	1 000 000
Detection limit	500	$-$	5 000 000
T(pyrolysis)	500	$-$	$-$
T(atomization)	1600	$-$	$-$

Comments: Argon-flushed optics required, with (preferably) a vacuum spectrometer. A detection limit of 500 pg has been obtained by emission from the S_2 species at 394 nm using a carbon rod atomizer with a hydrogen diffusion flame.[139] Sulphur has been determined indirectly, after separation of barium sulphate, by determining Ba.[144a]

Tellurium

λ(resonance): 214.3 nm
λ(background): Zn, 212.5 nm and Sb, 217.9 nm
Stock solutions: aqueous $-$ Te in the minimum quantity of HNO_3-HCl.
       organic $-$

	Graphite furnace	Tantalum filament	Flame
Sensitivity	100	600 (*a*)	500 000
Detection limit	50	300	100 000
T(pyrolysis)	400	400	$-$
T(atomization)	2000	2000	$-$

(*a*) Enhanced sensitivity with H_2.
Comments: Detection limits are improved with the use of an electrodeless discharge lamp. Suppression by Ca, K, Mg, and Na.[98a] Enhancement by HCl, HNO_3, and H_2SO_4.[98a] Losses of Te at elevated pyrolysis temperatures can be reduced by the addition of Ni.[102] Tellurium has been determined by the generation of TeH_2 with $NaBH_4$ and passing the TeH_2 into a graphite furnace.[101]

Thallium

λ(resonance): 276.8 nm
λ(background): Pb, 280.1 nm
Stock solutions: aqueous $-$ $Tl(NO_3)_3$ in water.
       organic $-$

	Graphite furnace	Tantalum filament	Flame
Sensitivity	50	200	500 000
Detection limit	20	70	200 000
T(pyrolysis)	750	750	$-$
T(atomization)	2200	1400	$-$

Comments: Suppression by HCl, $HClO_4$, and NaCl.[145]

Tin

λ(resonance): 224.6 nm (286.3 nm)
λ(background): Cd, 226.5 nm (283.9 nm)
Stock solutions: aqueous — Sn in the minimum quantity of H_2SO_4.
                organic — dibutyl-Sn-bis(2-ethylhexanoate) in xylene or
                MIBK.

	Graphite furnace	Tantalum filament	Flame
Sensitivity	100 (a)	400	1 000 000
Detection limit	100	100	100 000
T(pyrolysis)	1000	1000	—
T(atomization)	2500	2400	—

(a) Enhanced sensitivity with CH_4.

Comments: Detection limits are improved with the use of an electrodeless discharge lamp. Suppression by Cr[97] and H_2SO_4.[146] Enhancement by $HClO_4$ [146] and Ca.[80] Tin has been determined by the generation of SnH_4 with $NaBH_4$ and passing the SnH_4 into a graphite furnace.[101]

Titanium

λ(resonance): 364.3 nm, 365.4 nm
λ(background): Ni, 362.5 nm
Stock solutions: aqueous — TiO_2 in the minimum quantity of H_2SO_4—
                $(NH_4)_2SO_4$.
                organic — tetrabutyl titanate in butanol.

	Graphite furnace	Tantalum filament	Flame
Sensitivity	500 (a)	4000 (b)	2 000 000
Detection limit	500	1000	100 000
T(pyrolysis)	1300	1200	—
T(atomization)	2700	2300	—

(a) Enhanced sensitivity with CH_4. (b) Enhanced sensitivity with H_2.
Comments: Long atomization periods at high temperatures are essential to ensure complete removal of Ti from the atomizer. Although a nitrogen purge gas gives lower sensitivity than argon, it has been found to reduce memory effects with a graphite rod.[80]

Vanadium

λ(resonance): 318.4 nm, 318.5 nm
λ(background): Cu, 323.1 nm
Stock solutions: aqueous — V in the minimum quantity of HNO_3.
                organic — Bis(1-phenyl-1,3-butanediono)oxovanadium
                in xylene or MIBK.

	Graphite furnace	Tantalum filament	Flame
Sensitivity	200 (a)	1000 (b)	1 000 000
Detection limit	100	400	20 000
T(pyrolysis)	1600	1400	—
T(atomization)	2700	2300	—

(*a*) Enhanced sensitivity with CH_4. (*b*) Enhanced sensitivity with H_2
Comments: Long atomization periods at high temperatures are essential to
ensure complete removal of V from the atomizer. Enhancement by Al and
Ti.[147] Suppression by Ca and P.[147]

Zinc
λ(resonance): 213.9 nm
λ(background): 212.5 nm
Stock solutions: aqueous — Zn in the minimum quantity of H_2SO_4.
                 organic — Zn cyclohexanebutyrate in xylene or MIBK.

	Graphite furnace	*Tantalum filament*	*Flame*
Sensitivity	1	4	20 000
Detection limit	0.05	1	2 000
T(pyrolysis)	500	500	—
T(atomization)	2000	1200	—

Comments: Difficulties in determining Zn by electrothermal atomization
occur due to contamination (*e.g.* plastic sampling tips on micropipettes)
and obtaining sufficiently pure water. Suppression by Cl and PO_4.[71] Due
to the rapid rate of atomization for Zn, a recording system with a fast
response is required to avoid excessive curvature of the calibration graph.

6 Applications

Most published analytical methods utilizing flame atomic absorption spectrometry could be adapted very easily for use with electrothermal atomization. Some analytical methods which have been published using electrothermal atomization, however, would be more readily carried out using flame atomic absorption spectrometry. As a general rule, electrothermal atomization should not be used, in preference to the flame, for the determination of elements present in solid samples at concentrations greater than 1000 μg g^{-1} and in solutions at concentrations greater than 10 μg ml^{-1}.

In the following sections, published analytical methods have been included with sufficient information, whenever possible, to enable an analyst to carry out the determinations. The instrumental conditions used for each element can be obtained from Chapter 5. Since the majority of the analytical methods published are equally applicable to all types of electrothermal atomizer, albeit with different degrees of success, no attempt has been made to group the methods according to the type of atomizer used.

Where an application could be listed under more than one heading the method has been entered in the most appropriate section, *e.g.* meat under 'Food' but liver and kidney under 'Biological Tissues'. The Index should be consulted if in any doubt.

Within any one application section the matrices are given in alphabetical order and the elements determined are given in alphabetical order for each matrix. Where more than one element has been determined in an application the first element in the list determines the position of the entry in the table.

6.1 Oil and Oil Products

The concentrations of certain trace metals in crude oil and oil products are important for several reasons: (1) they can be used to help identify the source of oil spillages,[148] (2) they can poison catalysts, used in the processing of crude oil, (3) they may cause corrosion, when present in some fuel oils, (4) they are added to fuel and lubricating oils to improve performance, and (5) they can frequently be used to help in the diagnosis of engine wear.

Irrespective of the application, the use of electrothermal atomization is ideally suited to the direct analysis of oil products. Depending on the type of oil and the volatility of the element being determined, a pyrolysis temperature of 1000 °C is sufficient to remove the organic matrix of all but the very heavy oils; see Figure 6-1. When a sufficiently high pyrolysis

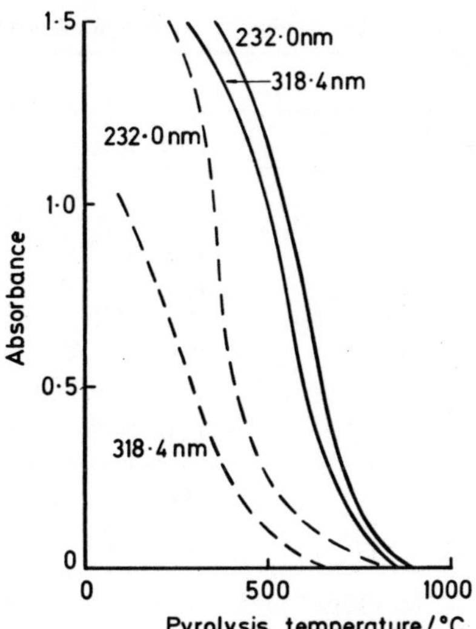

Figure 6-1   The variation of background absorbance with pyrolysis temperature for a light and a heavy fuel oil, at the absorption wavelengths for Ni and V.
–––: light fuel oil
——: heavy fuel oil, diluted 1 g to 10 ml with MIBK

temperature cannot be used then the application of background correction is essential to compensate for non-specific absorption/scatter signals.

Light oils can be injected directly to the atomizer while viscous oils only require dilution with xylene, MIBK, or THF prior to injection. All samples can be further diluted with the solvents if required. Calibration is carried out preferably with organometallic standards dissolved in the appropriate solvents or in a pure oil matrix, although aqueous standards are frequently satisfactory.

The main problem with the analysis of these materials, as with most organic liquids, is to ensure that the sample is retained reproducibly in the atomizer, and preventing it from irreproducibly soaking into graphite atomizers. This normally entails avoiding the use of large sample volumes. It may also be necessary, when analysing crude oils, to ensure that carbonaceous material does not build up in the atomizer.

Magnesium sulphonate has been recommended as an addition to samples prior to analysis, to ensure that all metals are converted into sulphonates, so as to prevent losses of the metals during pyrolysis.[149] The addition of up to 2 mg ml^{-1} of I_2 to samples being analysed for Pb has been proposed to overcome differences in response from various alkyl-lead compounds.[150,151]

A gas–liquid chromatograph fitted with an electrothermal atomizer detector has been used to determine the various alkyl-lead compounds in petroleum.[152]

Table 6.1   Determination of trace elements in oil and oil products

Element	References		
	Crude oil	Fuel oil	Lubricating oil
Ag	153, 154		153, 165–170
Al			165
Be		108	
Cd	154	160	
Co	154		
Cr		161	165, 169
Cu	153		153, 165–170
Fe	153, 155	161	153, 169, 170
Hg		161a*	
Mg	154		165, 170, 171
Mn		162	172
Na	154		
Ni	153, 155–157	161, 163	153, 165, 169, 170
Pb	153	131, 150–152	153, 165, 169, 170
Sn	154		169, 170, 173
V	155, 156, 158, 159	161, 163, 164	
Zn	154		

*After selective volatilization in a quartz tube and amalgamation with Au.

While the analysis of oils and oil products is reasonably straightforward, using the standard conditions for the elements outlined in Chapter 5, there have been many publications describing the determination of various elements. These are summarized in Table 6.1.

6.2 Metals

The presence of trace elements in ferrous and nonferrous metals can arise either from residual concentrations of these elements being carried through the production processes or as deliberate additions to the metals. In either case the concentrations of these elements are often critically important since they can severely alter the properties of the metals produced.

For most elements, concentrations greater than 100–500 μg g^{-1} are more conveniently and accurately determined by direct flame atomic absorption spectrometry, after sample dissolution. For lower concentrations, where separation and preconcentration techniques have hitherto been necessary, the use of electrothermal atomization enables the direct determination of elements down to 1 μg g^{-1}, or better, after sample dissolution. Analytical methods which involve the direct determination of elements in metals at concentrations below 500 μg g^{-1} are summarized in Table 6.2. Methods which involve the use of preconcentration techniques have been omitted where flame atomic absorption spectrometry would be the preferred method of analysis after the separation.

Table 6.2   Determination of trace elements in metals

Matrix	Element	Sample preparation	Reference
Chromium	Al		174
Copper	Al		174
	Ir, Pd, Pt, Rh	Dissolve in aqua regia; separate Ir, Pd, Pt, and Rh by ion exchange	125*a*
	Li	0.5 g + 10 ml HNO$_3$ (1 + 3); dilute to 50 ml. Determine by atomic emission	174*a*
	Pb	0.5 g + 10 ml HNO$_3$; dilute to 50 ml	129
	Se	Decompose in pressure vessel at 80 °C for 1 h with HClO$_4$ – HNO$_3$; evaporate to dryness; dissolve residue in 6M-HCl. Precipitate Se using Fe as collector; dissolve precipitate in HNO$_3$ and extract Se with 3,3′-diaminobenzidine	175
Gold	Fe	Dissolve in HNO$_3$ – HCl – H$_2$SO$_4$; precipitate Au with SO$_2$	126
Iron	Al	Dissolve 1 g in HCl–HNO$_3$; evaporate to dryness; redissolve in 6M-HCl; filter if necessary. Dilute to 100 ml with 6M-HCl. Extract Fe from an aliquot with MIBK; evaporate aqueous phase almost to dryness and then redissolve in 1M-HCl	176
	Al		174
	Al, Bi, Mg, Mn, Ni, Pb, V, Zn	0.5 g + 10 ml HNO$_3$; dilute to 50 ml	177, 177*a*
	As, Sb, Sn	1 g + 30 ml HNO$_3$ (1 + 4); dilute to 100 ml	97
	Bi	0.5 g + 7.5 ml HCl–HNO$_3$ (2 + 1); dilute to 50 ml	178
	Bi	1 g + 20 ml HNO$_3$ (1 + 1); dilute to 100 ml	178*a*
	Cr, Cu, Mn, Ni	Electrographic sampling	179
	Mn	Dissolve 0.1 g in HCl–HNO$_3$; evaporate to dryness. Redissolve in 0.02M-HCl	179*a*
	Pb	0.5 g + 10.5 ml HCl–HNO$_3$ – H$_2$O (10 + 5 + 6); dilute to 100 ml	180, 180*a*
	Sb	0.5 g + 7.5 ml HCl–HNO$_3$ (2 + 1); dilute to 500 ml	99

Table 6.2   *Continued*

Matrix	Element	Sample preparation	Reference
Mercury	Ag, Al, As, Au, Ca, Cd, Cr, Cu, Fe, Mn, Ni, Pb, Sb, V , Zn	Direct atomization from the metal or after dissolution in HNO_3	181
Molybdenum	Al	Direct atomization from the metal	182
Nickel	Al		174
	Bi, Pb, Se, Te, Tl	1 g + 30 ml $HNO_3 - HF-H_2O$ (1 + 1 + 1); evaporate to 5 ml; dilute to 50 ml	143
	Ir, Pd, Pt, Rh	Dissolve in aqua regia; separate Ir, Pd, Pt, and Rh by ion exchange	125a
Niobium	Al, Zn	Direct atomization from the metal	182
Silver	Fe	Dissolve in HNO_3; precipitate Ag with HCl	126
Tantalum	Al	Direct atomization from the metal	182
	Cr, Cu, Fe, Mn, Ni, Si	Direct atomization from the metal	183
Tin	Ag, Bi, Cd, Co, Cu, Ni, Pb	Dissolve in HCl	184
Tungsten	Al, Ca, Fe, Mo, Ni, Si	0.5 g + 30% H_2O_2	185
	Al, Zn	Direct atomization from the metal	182
	Se	0.5 g + 3 ml HF; add 1 ml HNO_3 slowly; dilute to 50 ml	146

For determinations in the presence of high matrix concentrations (greater than 1000 μg ml^{-1}), calibration graphs should be prepared in the presence of a similar matrix. Pyrolysis temperatures are usually not important in metal analysis, provided the temperature is sufficiently high to remove acids used in the sample dissolution. Selective volatilization of the matrix has only a limited use in these applications, *i.e.* the determination of non-volatile elements (*e.g.* V, Si, Mo) in volatile metal matrices (*e.g.* Zn, Cd, Hg). Background correction, at matrix concentrations below 10 000 μg ml^{-1}, is not normally essential, but the absence of background signals for a particular matrix should be confirmed.

6.3. Rocks, Minerals, and Soils

Geologists are usually concerned with the determination of relatively high concentrations of elements, and normally have access to large samples. Trace-element concentrations are increasing in importance, however, and in some cases only small samples are available, *e.g.* lunar and meteorite samples. Agronomists are frequently concerned with the levels of many trace elements in soils because of their effects on plant and, subsequently, animal life.

These materials are notoriously difficult to dissolve and usually require high volumes of acids or else fusion procedures to obtain complete dissolution of the samples. Because these methods of solution cause high blank values in analytical procedures there is a considerable advantage to be gained from the use of PTFE-lined high-pressure vessels; see Section 4.1.b. These considerably ease the problems of dissolving the sample and enable lower volumes of acids to be used. Knoop[186] has also described a special procedure for dissolving small silicate samples. Samples of 50 mg are introduced into plastic sealable microtest-tubes together with 0.5 ml HF; up to 10 of these sealed tubes are placed in a pressure vessel also containing HF. While the samples are being dissolved under pressure the HF in the pressure vessel ensures that the pressures inside and outside the sample tubes are equal. Sample aliquots can then be transferred directly from the sample tubes to the electrothermal atomizer. Care should be taken, however, when analysing HF solutions directly;[29] see Section 2.1.a.

Pyrolysis temperatures are not important provided they are high enough to remove acids used in the dissolution procedure. Background correction is usually required for samples containing high concentrations of the alkali metals and alkaline-earth elements. Calibration of analytical procedures normally entails a standard-additions method or the use of standards closely matching the sample matrix.

Analytical methods are summarized in Table 6.3.

Table 6.3   Determination of trace elements in rocks, minerals, and soils

Matrix	Element	Sample preparation	Reference
Carbonate rock	Co, Cr, Cu, Mo, Ni, V	0.5 g + 8 ml $HF-HClO_4$ (5 + 3); evaporate to dryness; repeat with $HF-HClO_4$. Add 5 ml $HClO_4$ and evaporate to dryness. Evaporate to dryness with HCl. Dissolve residue in HCl (1 + 9); filter and dilute to 100 ml with HCl (1 + 9)	187
	Pb	1 g + 10 ml HF; add 10 ml HNO_3 and evaporate to dryness. Dissolve residue in 10 ml HNO_3; dilute to 100 ml. For soluble Pb: 1 g + 10 ml HNO_3; filter and dilute to 100 ml	188, 189

Table 6.3   *Continued*

Matrix	Element	Sample preparation	Reference
Ilmenite	Cu	0.5 g + 10 ml HCl–HF (1 + 1); dilute to 50 ml	190
Iron ore	Co, Cr, Ni, Pb, Sb, V	0.2 g + 6 ml HF–HCl–HNO$_3$ (20 + 3 + 1) in a PTFE pressure vessel for 30 min at 150 °C. Cool, add H$_3$BO$_4$, and dilute to 100 ml	191
Sea-bottom sediments	Zn	Slurry 3 mg, finely ground sample, with 10 ml water. Transfer sample aliquot while stirring	192
Silicate rocks	Ag	Direct atomization from the solid	53
	Ag, Au	Leach twice with aqua regia; boil to dryness after each addition. For Au add a third aliquot of aqua regia; for Ag add an aliquot of HNO$_3$; heat to near boiling; cool, filter. Wash with 5% HCl for Au and 5% HNO$_3$ for Ag. Extract Au as the chloro-complex and Ag as the APDC complex into MIBK	193
	Ag, Cd, Tl	Vaporize impurities from solid samples in a graphite cup; collect impurities on a graphite rod; atomize directly from the rod	193a
	Ag, Cr, Cu, Pb, Zn	1 g + HF–HClO$_4$; dilute to 100 ml	122
	Al, Bi, Cd, Fe, Mg, Mn, Pb, Sr, Ti, Tl	10 mg + HF – HClO$_4$; evaporate HF and dilute to 25 ml. Some elements require separation for low level determinations	194
	Au, Co, Pb, V	Dissolve in HF–HClO$_4$; dilute to volume	195
	Ba	0.5 g + HF–HClO$_4$; evaporate to dryness. Dissolve residue in HCl	106a
	Be	0.2 g + HF–HClO$_4$; evaporate to dryness. Add 2–3 ml H$_2$SO$_4$; evaporate to dryness. Dissolve residue with 1.5 ml H$_2$SO$_4$; dilute to 50 ml	107
	Cd	Direct atomization from the solid or from the solid mixed with graphite (1 + 1)	52

Table 6.3　*Continued*

Matrix	Element	Sample preparation	Reference
Silicate rocks	Cd	Direct atomization from the solid	195a
	Cd, Co, Cu, Ni, Pb, Zn	1 g + 25 ml HF + 2 ml H_2SO_4; evaporate to fumes. Cool, add 7 ml HNO_3 + 21 ml HCl; evaporate to dryness. Add HNO_3 + H_2O; filter and dilute to 25 ml	196
	Co, Cu	Decompose with $HF-HNO_3$ in a pressure vessel; evaporate to dryness. Dissolve residue in H_2O; extract with DDTC into CCl_4	197
	Hg	0.5 g sample heated at 950 °C; Hg evolved amalgamated with Au. Redissolve Hg and electrolyse onto Au; atomise directly from Au electrode.	161a
	Te	1 g + 1 ml aqua regia + 15 ml HF; evaporate to dryness. Dissolve residue in 25 ml 6M-HCl; extract Te into MIBK	198
	Tl	0.2 g + 15 ml HF + 0.5 ml $HClO_4$; evaporate to dryness; convert into bromide salts by evaporation to dryness with HBr. Dissolve residue in 10–15 ml H_2O + 2 ml HBr; dilute to 25 ml. Extract with 5 ml isopropyl ether.	198a
	V	0.5 g + $HF-HClO_4$; evaporate to dryness. Add 2 ml HCl; evaporate to dryness. Dissolve residue in 20 ml 1M-HCl; dilute to 50 ml	147
Soil	Cd	10 g mixed with 50 ml 1M-HCl for 1 h. Centrifuge; neutralize to pH 4 with NaOH; extract Cd with APDC into MIBK.	199
	Co	Shake 2 g air-dry soil with 80 ml 2.5% MeCOOH overnight; filter and dilute to 100 ml. Transfer a 2 ml aliquot to a solution containing 12% pyridine and 5% KCNS; extract into 0.2 ml MIBK	199a

6.4 Waters

The analysis of all types of water for trace elements at extremely low levels has become of major importance, due mainly to the concern with environmental pollution. More research papers have been published covering the apparently simple task of analysing water than for any other single matrix. Only the major components of natural waters (Ca, K, Li, Mg, Na) are present at concentrations which allow direct determination by flame atomic absorption spectrometry. The increased sensitivity of electrothermal atomization has therefore been of immense importance in simplifying the current flame procedures, which require separation and preconcentration techniques, for the determination of trace elements.

For most elements in a water matrix (other than saline waters) direct injection of the sample to the atomizer can be used. Background correction and standard-additions calibration techniques are essential, however, for all but the purest water samples. The direct injection of saline water samples is generally not feasible unless certain methods of analysis are used; see below. The high NaCl content of these samples vaporizes during the analysis to cause (*a*) molecular absorption and light-scattering effects which are too great to compensate for, even with background-correction facilities, and (*b*) irreproducible results with graphite furnaces due to condensation of the NaCl at the electrical contacts between the graphite tube and the support cones. Segar and Gonzalez,[120] who studied the problems of background absorption with sea-water samples, using a graphite furnace, showed that the background signal could be compensated for all sea-water components except for NaCl and Na_2SO_4. If the maximum sample volume used was 20 μl, a pyrolysis temperature of 1250 °C for 25 seconds was required to remove the majority of these salts prior to atomization and thereby enable determinations to be carried out using background correction. Reference to Chapter 5 shows that this severely limits the number of elements which can be determined directly in sea-water. Segar and Gonzalez[120] also found it impossible to atomize volatile elements (Ag, Cd, Pb, Zn) prior to volatilization of the salt matrix of sea-water. Lundgren, Lundmark, and Johannson,[200] however, have shown that, by using an electrothermal atomizer which is designed so that the desired temperature is rapidly attained, it is possible to atomize and determine Cd in sea-water samples before the NaCl matrix volatilizes and interferes.

The problem of high background signals with sea-water samples has been reduced considerably in two ways.[201] (*a*) Addition of a concentrated solution of NH_4NO_3 to the sample converts NaCl, on heating, into a mixture of $NaNO_3$ and NH_4Cl. The NH_4Cl can be removed at temperatures below 500 °C, resulting in lower background signals. (*b*) Use of the reversed gas-flow system of the HGA 74, [see Section 2.1.a] enables higher matrix concentrations to be tolerated, with reduced background absorption.

The problems of correct sample collection and storage are particularly

Table 6.4   Determination of trace elements in waters

Matrix	Element	Sample preparation	Reference
Boiler feed	Cu, Fe, Ni	Direct injection	203
	Na	Direct injection	203*a*
Fresh	Ag, Cd, In, Pb	Direct injection (Cd) and solvent extraction. Dithizone–CCl_4 (Ag); dithizone–$CHCl_3$ (Pb); oxine–$CHCl_3$ (In)	204
	Ag, Cd, Mn, Pb, Zn	Direct injection	205
	Al, As, Cd, Co, Cr, Cu, Fe, Mn, Ni, Pb, Sn, Zn	Direct injection	206
	Al, Cd, Co, Cu, Mn, Ni, Pb	Direct injection (Al, Co, Cu, Mn, Ni) and extraction with 2% APDC into MIBK (Cd, Co, Cu, Ni, Pb)	207
	Al, Cr, Cu, Fe, Mn, Pb	Direct injection	208, 209
	As	500 ml sample acidified with 5 ml HNO_3 on collection. Direct injection	209*a*
	As, Sb	Direct injection	96
	As, Sb, Se	Direct injection after addition of Hg to reduce interferences in the determination of Sb	98
	As, Sb, Se, Te	Direct injection after 1 + 1 dilution with standards	98*a*
	Be	Direct injection	210
	Be	Acidify 1 l sample with 10 ml HCl. Add 3 g edta, adjust to pH 7 with NaOH. Extract Be with 2 ml acetylacetone into 30 ml $CHCl_3$. Be recovered into 6M-HCl using an ion-exchange column	210*a*
	Cd	Direct injection	111, 112
	Cd, Co, Cr, Cu, Fe, Mn, Mo, Ni, Pb	Direct injection	211
	Cd, Co, Cu, Ni, Pb	Direct injection	212
	Cd, Cr, Cu, Mn, Pb	Direct injection and solvent extraction	118
	Cd, Cu, Pb	Direct injection; addition of 0.004% edta enhances sensitivity	133
	Cd, Cu, Pb	Extraction with dithizone–CCl_4	130
	Cd, Pb (organically bound)	Filter particulates $\geqslant$0.15 μm. Separate organic acids (0.018–0.15 μm) by ultra-filtration; remove from filter with 100 ml 0.005M-NaOH. Add NH_4NO_3 and 1% HNO_3 to final volume	213
	Cu, Pb	Direct injection	123

Table 6.4   *Continued*

Matrix	Element	Sample preparation	Reference
Fresh	Hg	Separate by electrolysis onto Au; atomize directly from Au electrode	161a
	Pb	Direct injection after addition of 1% m/V ascorbic acid	134a
	Se	100 ml + 3 ml HNO_3 + 5 ml 30% H_2O_2; reduce volume to less than 50 ml by heating at 95 °C; dilute to 50 ml. Dilute a 5 ml aliquot to 10 ml, adding 0.1% Ni (as NO_3)	142
High purity	Ag, Cd, Co, Cu, Fe, Mn, Na, Ni, Pb, Si, Zn	Direct injection	214
	Al, Ca, Cd, Cr, Cu, Fe, Mg, Mn, Na, Pb, Si, Zn	Direct injection	215
	Co, Cr, Cu, Fe, Mn, Ni, Pb	Direct injection using automatic sampling system	61, 216
Rain/snow	Ag	100 ml melted snow acidified with 1 ml 0.1M-HNO_3. Extract with dithizone into CCl_4	217
	Ag, Cd, In, Pb	Direct injection (Cd) and solvent extraction. Dithizone–CCl_4 (Ag); dithizone–$CHCl_3$ (Pb); oxine–$CHCl_3$ (In)	204
	Ca	Direct injection	218
Sea	Ag, Cd, Co, Cr, Cu, Fe, Mn, Ni, Pb, V, Zn	Direct injection	120
	Be	Acidify 1 l sample with 10 ml HCl. Add 3 g edta, adjust to pH 7 with NaOH. Extract Be with 2 ml acetylacetone into 30 ml $CHCl_3$. Be recovered into 6M-HCl using an ion-exchange column	210a
	Cd	Neutralize to pH 8 with NH_3; extract Cd with NH_4DDTC into MIBK	219
	Cd, Cu, Fe, Ni, Pb, Zn	Direct injection (Cu, Fe, Ni) and separation with APDC–MIBK	220
	Cd, Cu, Pb	Extraction with dithizone into CCl_4	130
	Cd, Cu, Pb	Separate on Chetex 100 above pH 5; elute from resin with 50 ml 2M-HNO_3	221
	Cd, Cu, Pb, Zn	Filter; extract with dithizone–CCl_4	222
	Cd, Fe	Dilute	223

Table 6.4   *Continued*

Matrix	Element	Sample preparation	Reference
Sea	Cr	Boil with 0.1% edta	115
	Cu	Extract with DDTC into MIBK	224
	Cu	Direct injection or addition of NH_4NO_3	201
	Cu, Fe	1 ml + 0.1 ml 0.01M-HCl + 0.05 ml 2% APDC; extract into 1 ml MIBK	225
	Cu, Fe, Mn, Zn	Pass through hydrated Sb_2O_5 to remove Na salts	226
	Cu, Ni	Oxidize with $(NH_4)_2S_2O_8$; pass through Chitosan. Elute with 2% 1,10-phenanthroline in ethanol–H_2O or 1M-H_2SO_4	227
	Mo	Acidify to pH 2.5; pass through Chitosan; remove Mo with 1M-$(NH_4)_2CO_3$	228
	Pb	Extract with DDTC into DIBK	128
	Pb	Direct injection using T-shaped atomizer	229
	V	Acidify to pH 4; pass through Chitosan. Homogenize the resin and atomize V directly from a sample of the resin	73
Waste	Ag, Al, Cd, Cr, Cu, Fe, Mn, Ni, Pb, V, Zn	Direct injection	230
	As, Cd, Cr, Cu, Fe, Mn, Ni, Pb, Se	Filter; acidify to 0.2% HNO_3	231
	As, Sb	Direct injection	96
	As, Sb, Se, Te	Direct injection after 1 + 1 dilution with standards	98*a*
	Cd	Direct injection	111
	Cr	Acidify then direct injection	84
	Cu, Pb	Direct injection	123
	Se	Oxidize 500 ml sample with 5 ml HNO_3; add 2 ml $HClO_4$ and reduce volume to less than 50 ml. Add 3 ml HNO_3 and dilute to 50 ml	232
	Se	Evaporate sample to dryness; oxidize with HNO_3. Dissolve residue in 10 ml HCl (1 + 3); pass through IR-124 and dilute eluant to 50 ml	141
	Se	100 ml + 3 ml HNO_3 + 5 ml 30% H_2O_2; reduce volume to less than 50 ml by heating at 95 °C; dilute to 50 ml. Dilute 5 ml aliquot to 10 ml, adding 1% Ni (as NO_3)	142

important since the naturally occurring concentrations of most elements are extremely low; see Section 4.1.b.

Ediger has written a review of water analysis by atomic absorption spectrometry[202] which includes some useful information on electrothermal atomization methods.

Analytical methods published, for the analysis of water samples, are summarized in Table 6.4: many of the methods described can also be used for the determination of trace elements in mineral acids.

Table 6.5   Determination of trace elements in plants

Element	Sample preparation	Reference
Al, Cd, Cr, Cu, Fe, Mn, Pb, V	0.1 g + 0.1 ml H_2SO_4; heat to fuming then add 50% H_2O_2 dropwise until a clear solution is obtained; dilute to 10 ml	234
Be	Direct introduction of sample or wet ash 1 g with 25 ml $HNO_3-HF-H_2SO_4$ (2 + 2 + 1); heat to fumes and add 1 ml $HClO_4$ to complete oxidation; dilute to 100 ml	235
Cd	Dry ash 2 g at 450 °C; dissolve ash in 1M-HNO_3	236
Cd	Slurry 5 g dried sample with H_2O	110
Cd	0.1 g + 1 ml HNO_3-HClO_4 (2 + 1); heat at 60 °C for 15 min and then at 120 °C for 75 min; dilute to volume	237
Cd, Cr, Cu, Fe, Mn	Digest 0.25 g with 10 ml HNO_3 for 10 min; add 2 ml $HClO_4$ and evaporate to moist salts. Dissolve residue in 2 ml HNO_3 and dilute to 100 ml	238
Co	Ash at 450 °C; dissolve residue in HNO_3; adjust to pH 3−5 and extract with 1-nitroso-2-naphthol into 4 ml $CHCl_3$	121
Co	4 g dried sample + 20 ml 0.1M-HCl; boil; filter. Extract Co with 2 ml 2% APDC into $C_2H_2Cl_2$	239
Co	0.5 g + 17.5 ml HNO_3-HClO_4 (6 + 1); evaporate to 1−2 ml; add 5 ml H_2O. Cool; add 1 ml 40% Na citrate, 1 drop bromothymol blue, 1 drop methyl red and adjust to pH 5.3−5.7 with NH_3. Add 1 ml 30% H_2O_2 and 0.3 ml 1% 2-nitroso-1-naphthol in ethanol. Dilute to 12 ml and extract into 1 ml isoamyl acetate	240
Cu	50 mg + $HClO_4-HNO_3$ (1 + 9); evaporate to fumes. Dilute to 10 ml with 3% $HClO_4$ and extract with APDC into 1 ml MIBK	241
Cu, Fe, Mn	Direct introduction of sample	242
Mo	Direct introduction of sample	243
Mo	Ash 2 g dried sample at 500 °C for 5 h. Dissolve the ash in 20 ml HCl (1 + 19); dilute	244
Pb	Slurry 20 mg dried, ground sample with 2−8 ml water. Remove aliquot from slurry while stirring	245

6.5 Plants

Trace-element concentrations in plant materials are important guides to deficiencies in soil composition, and they can be used to select the correct type of fertilizer required to produce normal healthy plant growth. The concentrations of some elements are also important because many plants enter the human food-chain.

Apart from the increased sensitivity attainable with electrothermal atomization, the main advantages are (*a*) the possibility of directly analysing small plant samples, (*b*) readily determining the distribution of trace elements throughout the plant structure, and (*c*) because only small samples are required, a sample may be taken from a plant without significantly affecting its growth.

Background correction and standard-additions calibration graphs are usually required due to high levels of the alkali metals and alkaline-earth elements in plant materials. The nitric acid vapour-phase oxidation method proposed by Thomas and Smythe[233] should be particularly useful for minimizing blank values in sample preparation.

Published analytical methods are summarized in Table 6.5.

6.6 Food and Drugs

Certain trace elements (*e.g.* Cu, Fe, Mn) are an essential part of any balanced diet, while other elements (*e.g.* Cd, Pb) are extremely toxic to man. In both cases, careful control of the concentrations of these elements is of considerable importance. Electrothermal atomization could provide a very sensitive and simple technique for monitoring foods and drugs for these and many other elements. Despite the potential applications, there have been no reported methods for drug analysis, and a disappointingly small number of methods for foods.

Background correction is invariably required for methods where the sample matrix cannot be completely removed prior to atomization. Aqueous standards can be used for many matrices.

Analytical methods are summarized in Table 6.6.

6.7 Biological Fluids

The analysis of biological fluids represents the most important area of application of electrothermal atomization. Analyses can be carried out not only for the essential (Co, Cr, Cu, Fe, Mn, Zn) and toxic (Au, Be, Cd, Pb, Tl) trace elements but also for many others.

Direct injection or straightforward sample dilution is normally sufficient for analysis. In most cases aqueous standards are adequate for calibration; although care should be taken with urine samples, where matrix composition varies significantly. The highest possible pyrolysis temperature should be used that is compatible with the element being determined; see Chapter 5. Background correction is still required, however, with most direct-injection procedures.

Table 6.6   Determination of trace elements in foods

Matrix	Element	Sample preparation	Reference
Alcoholic drink	Cd	Direct injection + equivalent volume of HNO_3 (1 + 1)	110
Baby food	Cd	0.2 g + 5 ml HNO_3; dilute to 10 ml	110
Butter	Pb	1 g dissolved in 10 ml light petroleum benzine. Add 5 ml HNO_3 (1 + 1) and shake; pass through phase-separating filter paper. Evaporate aqueous phase to dryness then dissolve residue in 5 ml $HNO_3 - H_2O -$acetone (1 + 1 + 1)	246
Cheese	Cd	1 g + 10 ml NH_3 (1 + 9)	110
	Cu	1 g + 20 ml NH_3 (1 + 3) + 5 ml MIBK; heat at 60–65 °C for 10 minutes.	247
Coffee Powder	Pb	5 g dissolved in 50 ml H_2O; add 20 ml NH_4 citrate (50% m/V citric acid adjusted to pH 9 with NH_3); dilute to 100 ml	248
Fish/shellfish	Cd	Wet ash with HNO_3	223
	Cu	Wet ash with $HNO_3 - HClO_4$; evaporate to dryness; dissolve residue in 25 ml 0.1M-HCl	224
	Pb	1 g + 12 ml $HNO_3 - HClO_4$ (7 + 5); reduce volume to 1 ml; dilute to 100 ml.	249
	V	1 g dried sample + 5 ml $HNO_3 - HClO_4$ (4 + 1); evaporate to moist salts; dissolve residue in 1 ml HNO_3; dilute to 25 ml	146
Grain	Cd	Dry ash 2 g at 450 °C; dissolve residue in 1M-HNO_3	236
	Cd	Slurry 5 g with 20 ml H_2O	110
	Hg	Oxidize sample in combustion tube; collect Hg evolved on Au-plated carbon cup atomizer	250
Meat	Cd, Co, Cu, Mn, Ni, Pb	1 g + 10 ml HNO_3 (1 + 1); digest at 80 °C then reduce volume to 3 ml; filter; dilute to 25 ml	250a
Milk	Cd	Direct injection with equivalent volume of HNO_3 (1 + 1) added to sample in atomizer	110
	Co, Cu, Fe, Mn, Sr	Direct injection	251
	Cu	Direct injection	252
	Cu, Sr	Direct injection	92
	Pb	Direct injection	253

Table 6.6   *Continued*

Matrix	Element	Sample preparation	Reference
	Pb	Dry and partially ash sample after adding $Mg(NO_3)_2$; add HNO_3 and evaporate to dryness. Dissolve residue in HNO_3 (1 + 4); dilute	254
	Pb	12.5 ml + 1 ml HNO_3; evaporate to dryness; ash residue at 480°C for 3 h. Warm residue with 2 ml HNO_3 (1 + 1); add 10 ml H_2O and heat for 30 min. Filter and dilute to 25 ml	246
	Pb	Decompose with HNO_3 in pressure vessel for 2 h at 140 °C. Cool; dilute to 10 ml	255
Oils	Ca, Cd, Cr, Cu, Fe, K, Mg, Mn, Na, Ni, Pb, Zn	Direct injection	256
	Cu, Fe, Ni	Dissolve oil in isoamyl acetate. Introduce O_2 to the purge gas during pyrolysis stage to aid decomposition of sample	86
Sugar	Cr	1 g + 10 ml H_2O (unrefined sugars centrifuged to remove insoluble fraction), dry and ash 1 ml aliquot (together with a portion of the insoluble fraction) in an O_2-plasma low-temperature asher. Alternatively, for unrefined sugars, ash 0.1 g directly. Dissolve residue in 1 ml 1.2M-HCl	257
Synthetic protein	Cd, Pb	1 g, dried sample, +5 ml HNO_3– $HClO_4$ (4 + 1); evaporate to moist salts; dissolve residue in 1 ml HNO_3 and dilute to 25 ml	146
Tea powder	Pb	5 g + 50 ml H_2O; add 20 ml NH_4 citrate (50% m/V citric acid adjusted to pH 9 with NH_3); dilute to 100 ml	248

Because direct-injection procedures are normally adequate, sample volume requirements are minimal, *e.g.* 5–50 μl. This is particularly important for the analysis of blood samples from young children.

Analytical methods are summarized in Table 6.7. As many of the methods for whole blood (B), plasma (P), and serum (S) are similar, they have been grouped together, and identified in the reference column by the appropriate code letter.

Table 6.7   Determination of trace elements in biological fluids

Matrix	Element	Sample preparation	Reference
Blood/ plasma/ serum	Al	Direct injection	258 (S)
	Al, Co, Cr, Cu, Fe, Mn, Li, Pb	Dilute 1 + 4 with H_2O	259 (S)
	Au	Direct injection	260 (S)
	Au, Co, Li	Direct injection	59 (P)
	Cd	Direct injection	261 (S)
	Cd	1 ml + 7 ml acetone; add 1 ml $HClO_4$; dilute to 10 ml with acetone	262 (B)
	Cd	Extract into 0.2 ml MIBK	263 (B)
	Cd	Dilute 0.1 ml to 5 ml	264 (B)
	Cd	0.5 ml + $HNO_3 - H_2O_2$; evaporate to dryness; dissolve residue in 1% HNO_3	265 (B)
	Cd	Spot blood onto filter paper; cut a 4 mm diameter disc from the paper; analyse the disc directly	266 (B)
	Cd	0.1 ml + 0.9 ml H_2O; add 20 μl heparin and 20 μl 2% Triton X-100	266a (B)
	Cd, Cu, Fe, Mg, Mn, Pb, Zn	Freeze dry; low-temperature-plasma ash then wet ash with HNO_3	267 (B)
	Cd, Pb	Direct injection together with twice the volume of HNO_3 (Pb) or three times the volume of HF (Cd). For Cd, results improved by allowing O_2 to enter atomizer during pyrolysis stage	114 (B)
	Cd, Pb, Tl, Zn	Ash in an h.f. induction furnace	268 (B)
	Co, Cu, Mn	Dilute with water. For Co in serum, dry at 750 °C for 5 min and atomize ash directly	269 (B, S)
	Cr	0.2 ml + 0.4 ml $HClO_4$ in screw-top container; heat at 135 °C for 45 min; cool; add 0.1 ml H_2O_2 and heat at 135 °C for 15 min then reflux at 180 °C for 45 min. Evaporate to dryness; dissolve residue in 0.4 ml 0.3M-HCl	119 (P)
	Cr	1 + 1 dilution with H_2O	115 (B)
	Cr	Direct injection	270 (S)
	Cr	Direct injection together with half the sample volume of both 5M-HNO_3 and 30% H_2O_2	271 (S)
	Cr, Mn	Direct injection	272 (S)

Table 6.7 *Continued*

Matrix	Element	Sample preparation	Reference
Blood/ plasma/ serum	Cu	Direct injection	273 (S)
	Cu	Dilute 1 + 9 with 10^{-2} M-HNO$_3$	274 (S)
	Cu, Fe, Mg, Pb, Zn	Direct injection	275 (B, P)
	Cu, Zn	Separate plasma protein fractions on cellulose acetate using electrophoresis. Cut fractions from cellulose acetate and analyse strips directly	276 (P), 277 (P)
	Fe	Direct injection	278 (S), 279 (S), 280 (S)
	Mg	Dilute	171 (S), 281 (S)
	Mn	Dilute	282 (S)
	Mn, Sr	Direct injection	283 (S), 284 (S)
	Ni	Direct injection together with HNO$_3$ to aid oxidation	285 (S)
	Ni	Evaporate 3 ml to dryness; ash at 560 °C for 5 h. Dissolve the residue in 3 ml 1M-HCl; evaporate to near dryness. Add 3 ml H$_2$O, 2 drops 10^{-5}% (NH$_4$)$_2$H citrate, and 0.2 ml of 1% DMG. Adjust to pH 9 with NH$_3$ and extract with 2 × 1 ml MIBK	286 (S)
	Pb	Direct injection	81 (S), 196 (B), 229 (B), 287 (B), 288 (B), 289 (B), 290 (B), 291 (P)
	Pb	Direct injection; dry; add twice sample volume of 30% H$_2$O$_2$	292 (S)
	Pb	1 + 10 dilution with water	293 (B)
	Pb	Dilution with Triton X-100	294 (B), 295 (B), 296 (B), 297 (B).
	Pb	0.1 ml + drop Sopanim, 0.2 ml formamide, 0.1 ml 2% APDC. Extract into 0.5 ml MIBK	298 (B)
	Pb	50 μl + 50 μl HNO$_3$; heat at 70 °C	299 (B)
	Pb	Spot blood onto filter paper; cut a 4 mm diameter disc from paper; analyse disc directly	300 (B)

Table 6.7   *Continued*

Matrix	Element	Sample preparation	Reference
Blood/ plasma/ serum	Pb, Tl, Zn	Direct injection; addition of HNO_3 to the sample in the atomizer aids oxidation	301 (B)
	Pt	Dilute 1 + 1 with H_2O	125a (B)
	Zn	Direct injection	302 (S)
	Zn	1 + 9 dilution with water	303 (P)
Cerebrospinal fluid	Mn	Dilute	282
Sweat	Pb	Direct injection	304
Urine	Cd	Direct injection	263
	Cd	0.1 ml diluted to 5 ml	264
	Cd	0.5 ml + HNO_3 – H_2O_2; evaporate to dryness. Dissolve residue in 1% HNO_3	265
	Cd	Stabilize samples with 0.1 g edta in 100 ml. Acidify with HNO_3 to pH 2; separate Cd by electrolysis onto Pt wire. Atomize directly from wire	76a
	Cd, Pb, Tl	Extract with NaDDTC at pH 7 into 1 ml MIBK. Add $CaCl_2$ to overcome presence of edta	305
	Cd, Pb, Tl, Zn	Digest with HNO_3	268
	Cr	Direct injection	306, 307
	Cr	0.2 ml + 0.4 ml $HClO_4$ in screw-top container; heat at 135 °C for 45 min. Cool; add 0.1 ml H_2O_2; heat at 135 °C for 15 min then reflux at 180 °C for 45 min. Evaporate to dryness; dissolve residue in 0.4 ml 0.3 M-HCl	119
	Mn	Dilute with H_2O (1 + 1)	282
	Pb	Direct injection	229
	Pb	Dilute	81
	Pb	0.5 ml + 0.5 ml 0.1M-edta (enhances sensitivity) + 0.5 ml H_2O	134
	Pb, Tl, Zn	Direct injection; addition of HNO_3 to the sample in the atomizer aids oxidation	301
	Tl	1 ml diluted to 10 ml with 1% H_2SO_4	145
	Zn	Direct injection	302

6.8 Biological Tissues

As with biological fluids, the use of electrothermal atomization is potentially a very useful analytical technique for the analysis of biological tissues. Analysis of tissues is, however, of less importance than the analysis of fluids, this type of analysis being most often required for *post mortem* studies or for laboratory experimentation with animals. Although direct analysis of solid samples is feasible, it is more normal to carry out a wet oxidation of the sample or to use one of the commercial solvents designed for the solubilization of tissue material.

Background correction is often required during analyses, but aqueous standards are normally adequate where a wet oxidation procedure has been used for sample preparation.

Published analytical methods are summarized in Table 6.8.

Table 6.8   Determination of trace elements in biological tissues

Matrix	Element	Sample preparation	Reference
Bone	Cd, Cu, Mn, Pb, Zn	1 g + 2 ml 25% TMAH in ethanol; separate soluble fraction and dissolve inorganic residue in HNO_3	308
Cephalin	Ca	20 mg + 1 ml C_6H_6 –$CHCl_3$ – MeOH (3 + 2 + 1)	309
Eggs	Cd	2 g + 5 ml Triton X-100 (1 + 19)	110
Faeces	Pb	0.5 g + 20 ml HNO_3; reflux for 1 h; dilute to 50 ml	132
Finger/ toenails	Cd	10 mg + 1 ml 8M-HNO_3 + 0.5 ml 10% H_2O_2; dilute to volume	263
	Cd, Cu, Mn, Pb, Zn	1 g + 2 ml 25% TMAH in alcohol; dilute to 10 ml with H_2O	308
	Cu	Atomization from solid sample	310, 311
Hair	Cd	10 mg + 1 ml 8M-HNO_3 + 0.5 ml 10% H_2O_2; dilute to volume	263
	Cd, Cu, Mn, Pb, Zn	1 g + 2 ml 25% TMAH in alcohol; dilute to 10 ml with H_2O	308
	Pb	Atomization from solid sample	312
Soft tissue	Ag, Cu, Mn, Pb	Atomization from dried sample	313
	Al, V	30 mg dried sample ashed at 650 °C for 18 h; dissolve residue in 50 μl HNO_3 (1 + 19), dilute to 1 ml. Alternatively, grind and sieve (200 mesh) sample and analyse solid sample directly	313*a*

Table 6.8   *Continued*

Matrix	Elements	Sample preparation	Reference
Soft tissue	Be	Atomization from dried sample	235
	Cd	5 mg dried sample + 0.5 ml $HNO_3-H_2SO_4$ (1 + 4); heat at 90 °C in a sealed plastic tube, overnight. Add 1 ml of H_2O; further dilute 50 μl aliquot to 1 ml	314
	Cd	0.2 g dried sample + 5 ml HNO_3; dilute to 10 ml	110
	Cd	0.1 g + 1 ml $HNO_3-H_2SO_4-HClO_4$ (3 + 1 + 1); digest for 2 h at 74 °C. Dilute to 25 ml	314*a*
	Cd	10 mg + 1 ml HNO_3 + 1 ml 10% H_2O_2; dilute	263
	Cd	10 mg dried sample + 5 ml H_2SO_4; heat gently. After 5 min add HNO_3 dropwise until reaction ceases; heat for 5 min, cool, and dilute to 100 ml	314*b*
	Cd, Cr, Cu, Fe, Mn, Pb, Zn	0.25 g dried sample + 10 ml HNO_3; heat for 10 min; add 2 ml $HClO_4$ and evaporate to moist salts; dilute to 10 ml	238
	Cd, Cu	Air dry 15 mg at 80 °C for 24 h; add 1 ml 1M-HNO_3 and digest at 80 °C for 24 h. Evaporate to dryness; dissolve residue in 2 ml 10^{-2} M-HNO_3	315
	Cd, Cu, Mn, Pb, Zn	1 g + 2 ml 25% TMAH in alcohol; dilute to 10 ml with H_2O	308
	Co, Mn, Zn	Atomization from solid sample	316
	Cr, Cu, Mn, Ni	Wet ash	285
	Cu, Mn	0.25 g + $H_2SO_4-H_2O_2$. For smaller samples: low-temperature ash and dissolve ash in 0.2–2 ml 10% HCl	317
	I	Dry 20–30 mg, under vacuum, at 50 °C; digest with 1 ml aqua regia; evaporate to near dryness. Add 0.5 ml HNO_3 and re-evaporate to near dryness. Dilute to 5 ml	318
	Mo	Atomization from dried sample	243
	Pb	5 g + 5 ml Soluene-350; heat at 55 °C for 4 h (for liver tissue heat for 24 h). Dilute to 10 ml with toluene	319

Table 6.8   *Continued*

Matrix	Elements	Sample preparation	Reference
Soft tissue	Pb	0.2 g digested with HNO_3. Decompose excess HNO_3 by dropwise addition of HCOOH	319*a*
	Pd, Pt	1 g dried sample + 5 ml HNO_3 (x 4) + 5–10 mg NaCl + 4 ml aqua regia (x 2) + 4 ml HCl (x 3); evaporate to dryness after each acid addition. Dilute to 10 ml with 20% HCl	320
	Se	1 g dried sample + 10 ml HNO_3: reduce volume to 3 ml by heating. Cool, add 6 ml $HClO_4$ and 3 ml H_2SO_4. Heat to fumes of SO_3. Add 1 ml 30% H_2O_2 and boil to fumes of SO_3. Repeat additions of H_2O_2 twice and finally heat for 5 min. Cool and dilute to 3 ml	143*a*
Teeth	Al	10 mg + 0.5 ml HNO_3 (1 + 5) in sealed plastic tube	95
	Cd, Cu, Mn, Pb, Zn	1 g + 2 ml 25% TMAH in ethanol; separate soluble fraction and dissolve inorganic residue in HNO_3	308
Yeast	Cr	50 mg + 250 μl HNO_3; heat at 160 °C for 3 h in a pressure vessel	320*a*

6.9   Air Particulates

Airborne dust is generated by winds removing particulate matter from the earth's surface or as a result of chemical reactions, either industrial or domestic. Particulate matter larger than 10 μm very rapidly settles out and returns to the earth's surface while smaller particles are dispersed by winds over considerable distances before they eventually return to the earth's surface by sedimentation or by the actions of rain and snow. The high sensitivity available with electrothermal atomization avoids the problem of large sample volumes (required for flame atomic absorption procedures) for the determination of the trace elements in air particulate matter.

Begnoche and Risby[321] have provided a good introduction to the problems of air sampling, emphasizing the difficulties caused by sample contamination, high blank values, removal of particles from the filter, and multi-element analyses. The preferred method of analysis is the use of organic or inorganic filters (of appropriate pore size, *e.g.* 0.2 μm) to collect the air particulates, followed by an acid dissolution of the sample. This technique is advantageous if several elements are to be determined or if the

Table 6.9   Determination of trace elements in atmospheric particulate matter.

Element	Sample preparation	Reference
Ag	AgI particles collected by passing air through a 1 μm pore-size nylon filter. Remove AgI by soaking in acetone	328
Ag, Be, Cd, Hg, Pb, Se	Direct filtration through a carbon tube, which is then used in the atomizer	329
Al, Ca, Cd, Co, Cr, Cu, Fe, Mg, Mn, Ni, Pb, Sn, Zn	Pass 20–300 1 air through a filter (0.05 or 0.10 μm pore size). Remove particles with 10 ml HCl or aqua regia (1 + 1), using ultrasonic agitation	321
Be	Pass air through a porous carbon tube, which is then used in the atomizer	330
Be	Pass air through glass-fibre filter; dissolve filter in 5 ml $HF-HNO_3$ (3 + 2). Evaporate to near dryness; add 2 ml H_2SO_4 and heat to fumes. Add 3 drops HNO_3 and heat to fumes; dilute to 100 ml	331
Cd	Pass 400 m^3 air through filter paper; use one-eighth sector of paper. Add 50 ml 0.1M-HNO_3 and remove particulate matter with ultrasonic agitation; repeat with 30 ml 0.1M-HNO_3; dilute to 100 ml	332
Cd	Pass air through glass-fibre filter; dissolve filter in $HF-HNO_3$. Evaporate to 0.5 ml; dilute to 10 ml	333
Cd	Pass 400 ml air through filter (0.22 μm) fitted in a porous graphite cup atomizer. Add 2 μl 0.1% H_3PO_4 to cup and atomize directly	113
Hg	Pass air through Au-plated carbon tube, which is then used in the atomizer	136, 137, 334, 335
Pb	Pass 10 1 air through filter paper; soak filters in 2 ml HNO_3 (1 + 1); warm. Dilute to 10 ml	336
Pb	Pass 500 ml through a porous carbon cup, which is then used in the atomizer	337
Pb	Pass 400–1200 m^3 air through filter paper; remove Pb from filter segment with 50 ml 0.1M-HNO_3, using ultrasonic agitation. Dilute to 100 ml	338
Pb	Pass 200 ml air through filter (0.22 μm) fitted in a graphite cup atomizer; add 2 μl 0.1% H_3PO_4 to cup and atomize directly	339, 339a
Pb	Pass air through a porous carbon cup, which is then used in the atomizer	340
Pb	Collect particulate matter, using a cascade impactor, on a tantalum foil, which is then used in the atomizer	341
Pb(alkyl)$_4$	Pass air through 15 ml 1M-ICl in 5% HCl; add 20 ml 1% NH_4 citrate–2% Na_2SO_3 in NH_3 (4 + 21) and 2 drops 0.1M-edta. Separate Pb with 5 ml 0.002% dithizone in CCl_4; back-extract Pb into 2 ml 1% $HNO_3-1\%$ H_2O_2	342
Pt	Leach air filters with 1M-HCl or 50% aqua regia for soluble or total Pt, respectively	125a

concentrations of an element vary widely. The direct sampling method whereby the air sample is pulled through a graphite cup or tube, which is then used as the atomizer, is an extremely rapid and sensitive technique when only a single element is to be determined, over a small concentration range.

Calibration of an analytical determination is readily carried out using aqueous standards; calibration of the overall procedure is considerably more difficult, owing to the lack of widely available standard contaminated air samples. Robinson and Wolcott, who, together with other co-workers, have reported several analytical methods for the determination of atmospheric pollutants using an r.f. induction-heated furnace,[322-326] investigated this problem. As a result of the work, they have published a very good review of the methods they used to produce standard contaminated air samples.[327] These authors concluded that direct injection of standard solutions to the atomizer was satisfactory as a calibration method.

Published analytical methods are summarized in Table 6.9.

6.10 Refractory Oxides and Related Materials

Refractory oxides present severe problems for analysis, by most analytical techniques. The main problem, with solution techniques, arises from the difficulty of dissolving samples without introducing high blank levels due to the large acid volumes or fusion mixtures required. For this reason the application of PTFE-lined pressure vessels is particularly useful.

The direct analysis of refractory materials, using electrothermal atomization, is very difficult due to irreproducible heating of the samples. However, any method which avoids the need to dissolve the sample is extremely valuable. The technique of forming a slurry and analysing the slurry directly could be used to great advantage in this area of application, provided that a method of forming a relatively stable slurry can be obtained.

The use of standard-additions calibration or a close matching of standards to sample matrix is normally required. Background correction is usually unnecessary except for sample solutions with a high solids content or high alkali-metal or alkaline-earth concentration.

Analytical methods are summarized in Table 6.10.

Table 6.10 Determination of trace elements in refractory oxides and related materials.

Matrix	Element	Sample preparation	Reference
Boric oxide	Co, Cr, Cu, Fe, Mn, Ni	0.5 g + 6 ml HF–HClO$_4$ (2 + 1); evaporate to dryness. Dissolve in 5 ml H$_2$O	146

Table 6.10   *Continued*

Matrix	Element	Sample preparation	Reference
Calcium carbonate	Ba	2.5 g + 15 ml 4M-HNO_3; dilute to 100 ml	105
	Co, Cr, Cu, Fe, Mn, Ni	1 g + 3 ml $HClO_4$ evaporate to dryness; add 0.1 ml $HClO_4$ + 4 ml H_2O. Add 5 ml 5% Na DDTC–10% Na acetate buffered at pH 6. Extract into 1 ml MIBK	343
Glass (alkali/ alkaline earth)	Co, Cr, Cu, Fe, Mn, Ni	1 g + 9 ml HF–$HClO_4$ (4 + 5); evaporate slowly to dryness; add 0.1 ml $HClO_4$ + 4 ml H_2O. Add 5 ml 5% Na DDTC–10% Na acetate buffered at pH 6. Extract into 1 ml MIBK	343
	Cu, Fe	Decompose 0.2 g with 10 ml HNO_3–HF (3 + 2), using vapour-phase attack. Dissolve residue in 2 ml H_2O; add 2 ml 0.2% NH_4 TDTC buffered at pH 6 with Na acetate. Extract into 0.2 ml $CHCl_3$	67
Glass (lead)	Co, Cr, Cu, Fe, Mn, Ni	1 g + 10 ml HF–H_2SO_4 (1 + 1); evaporate slowly to dryness. Break up residue and shake vigorously with 10 ml 0.1% HCl for 5 min. To a 5 ml aliquot add 5 ml 5% Na DDTC–10% Na acetate buffered at pH 6. Extract into 1 ml MIBK	344
Rare-earth oxides	Ag, Au, Ca, Cu, Fe, K, Mg, Mn, Na, Zn	Atomize directly from a graphite cup	345
Silica	Cu, Fe	1 g + 4 ml HF. (Highly calcined samples require the use of a PTFE-lined pressure vessel.)	29
	Ti	0.2 g + 5 ml H_2SO_4; add HF dropwise until sample completely dissolved. Heat to fumes; dilute to 100 ml	346
Sodium carbonate	Co, Cr, Cu, Fe, Mn, Ni	1 g + 3 ml $HClO_4$; evaporate to dryness; add 0.1 ml $HClO_4$ + 4 ml H_2O. Add 5 ml 5% Na DDTC–10% Na acetate buffered at pH 6. Extract into 1 ml MIBK	343
Titanium dioxide	Al, Cu, Fe, Mn	0.5 g + 5 ml HF; dilute to 50 ml	347, 348
	Cu	0.5 g + 10 ml HCl–HF (1 + 1); dilute to 50 ml	190

Table 6.10   *Continued*

Matrix	Element	Sample preparation	Reference
Titanium dioxide	Cu, Fe, Mn, Pb	Slurry 1 g with 100 ml H_2O containing 0.005% sodium hexametaphosphate as the dispersing agent	146
	Pb	Mix 100 mg with 1 g graphite; determine Pb directly in 5 mg of the mixture	56
	Sb	1 g + 10 ml H_2SO_4 (1 + 1); extract the chloride complex into 10 ml isopropyl ether	349
	V	Fuse 1.5 g with 3.0 g Na_2CO_3; boil crushed melt with H_2O; centrifuge and dilute to 25 ml	350
Uranium oxide	Co, Cr, Fe, Mn, Mo, V	0.5 g + 2–3 ml HNO_3 (1 + 1); evaporate to dryness. Dissolve residue in 0.2M-HNO_3 and dilute to 250 ml	351

6.11 Other Analytical Applications

Several useful analytical applications have been published which do not fit conveniently into any of the other categories. Of particular interest are the procedures for the direct analysis of plastic-type materials and the procedures for the determination of Ba and Sb in gunshot residues. These and other applications are given in Table 6.11.

Table 6.11   Determination of trace elements in miscellaneous materials.

Matrix	Element	Sample preparation	Reference
Bullets (lead)	Ag, Cu, Sb, Zn	Dissolve in 10% HNO_3	352
Catalysts	Pt	0.5 g + 10 ml aqua regia; heat at 95 °C for 90 min; evaporate to dryness. Dissolve residue in 7 ml HCl and dilute to 25 ml	353
Coal	Be	1 g + 25 ml HNO_3–HF–H_2SO_4 (2 + 2 + 1); evaporate to fumes; add 1 ml $HClO_4$ and dilute to 1 litre	235
	Hg	0.5 g sample heated at 950 °C; Hg evolved amalgamated with Au. Dissolve Hg and electrolyse onto Au; atomize directly from Au electrode	161a
Gallium arsenide	Cr, Te	Separate matrix by chromatography	354
	Zn	Dissolve in HNO_3–HCl	355

Table 6.11   *Continued*

Matrix	Element	Sample preparation	Reference
Gunshot residues	Ba	Collect samples with cotton swabs and leach with 1 ml 0.2% HNO_3	355a
	Ba, Pb, Sb	Collect samples with cloth moistened in dilute HCl and then leach with 10 ml 3M-HCl. To one half of the extract add thymol blue indicator and adjust to pH 2–3 with NH_3; extract Pb with 1% APDC into 0.5 ml MIBK. Add further thymol blue to the aqueous layer and adjust to pH 8–9 with NH_3; extract Ba with 0.1M-TTA in MIBK. To the second half of the extract add 200 μ1 0.05M-$Ce(SO_4)_2$; extract Sb into 0.5 ml MIBK	106
	Ba, Pb, Sb	Collect samples with cotton swabs moistened with 10% HCl. Leach with 10% HCl	356
	Ba, Sb	Collect samples with cotton swabs moistened with 5% HNO_3. Leach with 2 ml 1M-HNO_3	356a
Paper and pulp	Cu, Fe, Mn, Si	Atomization from solid sample	357
Plastics	Al, Cu, Fe	Atomization from solid sample	357
	Au	Atomization from solid sample	358
	Au	1–2 cm^2 of photographic film decomposed by enzyme action followed by HNO_3–H_2O_2 (1 + 1). Extract Au as the bromide into MIBK	359
	Cr, Cu, Fe	Atomization from solid sample or dissolve 1 g in 100 ml 0.16M-HNO_3 for water-soluble polymers	360
	Cu, Pb	Atomization from solid sample	229
Silicon	Cu, Fe	0.5 g + 4 ml HF; dilute to 50 ml	146, 348
Silicon nitride	Cu, Fe	0.5 g + 5 ml HF–HCl (1 + 1) in PTFE-lined pressure vessel; dilute to 50 ml	146, 348
Soap	Al, Cu, Zn	Dilute to 0.01 mol l^{-1}	361

6.12 Theoretical

The use of electrothermal atomizers for measuring fundamental parameters has many advantages which arise from the ease with which the vaporization and atomization processes can be controlled. The experimental variables, *e.g.* gas type and pressure and chemical reactions, can be controlled and varied as required with many designs of apparatus. King's initial work[1,2] with graphite furnaces for the measurement of emission spectra was a direct result of these advantages.

Other parameters which have been measured using electrothermal atomization include the following:

(i) activation energy for the removal of an element from a matrix[362,363]
(ii) atomic diffusion coefficients,[7,364,365]
(iii) dissociation energies,[362]
(iv) Lorentz broadening of the spectral line profile,[7,366]
(v) oscillator strengths,[7,367]
(vi) reduction of metal oxides by carbon,[363]
(vii) vacuum-u.v. spectroscopic studies,[368-370]
(viii) vapour-pressure measurements,[363,371,372]

References

1 King, A. S., *Astrophys. J.*, 1905, **21**, 236.
2 King, A. S., *Astrophys. J.*, 1908, **27**, 353.
3 Mandelshtam, S. L., Semenov, N. N., and Turovtseva, Z. M., *Zhur. analit. Khim.*, 1956, **11**, 9.
4 Zaidel, A. N., Kaliteevskii, N. I., Lipis, L. V., Chaika, M. P., and Belyaev, Y. I., *Zhur. analit. Khim.*, 1956, **11**, 21.
5 L'vov, B. V., *J. Eng. Phys. (U.S.S.R.)*, 1959, **2**, 44.
6 L'vov, B. V., and Lebedev, G. G., *Zhur. priklad. Spektroskopii*, 1967, 7, 264.
7 L'vov, B. V., *Spectrochim. Acta*, 1969, **24B**, 53.
8 Brandenburger, H., and Bader, H., *Atomic Absorption Newsletter*, 1967, **6**, 101.
9 Massmann, H., *Z. analyt. Chem.*, 1967, **225**, 203.
10 Massmann, H., *Spectrochim. Acta*, 1968, **23B**, 215.
11 Woodriff, R., and Ramelow, G., *Spectrochim. Acta*, 1968, **23B**, 665.
12 Woodriff, R., Stone, R. W., and Held, A. M., *Appl. Spectroscopy*, 1968, **22**, 408.
13 West, T. S., and Williams, X. K., *Analyt. Chim. Acta*, 1969, **45**, 27.
14 Alder, J. F., and West, T. S., *Analyt. Chim. Acta*, 1970, **51**, 365.
15 Bratzel, M. P., Dagnall, R. M., and Winefordner, J. D., *Analyt. Chim. Acta*, 1969, **48**, 197.
16 Donega, H. M., and Burgess, T. E., *Analyt. Chem.*, 1970, **42**, 1521.
17 Montaser, A., Goode, S. R., and Crouch, S. R., *Analyt. Chem.*, 1974, **46**, 599.
18 Katskov, D. A., Kruglikova, L. P., L'vov, B. V., and Polzik, L. K., *Zhur. priklad. Spektroskopii*, 1974, **20**, 739.
18a L'vov, B. V., *Talanta*, 1976, **23**, 109.
19 Robinson, J. W., and Wolcott, D. K., *Analyt. Chim. Acta*, 1975, **74**, 43.
20 Massmann, H., and Gucer, S., *Spectrochim. Acta*, 1974, **29B**, 283.
21 Molnar, C. J., Chuang, F. S., and Winefordner, J. D., *Spectrochim. Acta*, 1975, **30B**, 183.
22 Ottaway, J. M., and Shaw, F., *Analyst*, 1975, **100**, 438.
22a Epstein, M. S., Rains, T. C., and O'Haver, T. C., *Appl. Spectroscopy*, 1976, **30**, 324.
23 Segar, D. A., and Gonzalez, J. G., *Atomic Absorption Newsletter*, 1971, **10**, 94.
24 Kahn, H. L., and Slavin, S., *Atomic Absorption Newsletter*, 1971, **10**, 125.
24a Sperling, K. R., *Z. analyt. Chem.*, 1976, **279**, 205.
25 Fernandez, F. J., *Atomic Absorption Newsletter*, 1972, **11**, 123.
26 Issaq, H. J., and Zielinski, W. L., *Analyt. Chem.*, 1975, **47**, 2281.
26a Sperling, K. R., *Atomic Absorption Newsletter*, 1976, **15**, 1.
27 Kuzovlev, I. A., Kuznetsov, Y. N., and Sverdlina, O. A., *Zavodskaya Lab.*, 1973, **39**, 428.
28 Ortner, H. M., and Kantuscher, E., *Talanta*, 1975, **22**, 581.
28a Runnels, J. H., Merryfield, R., and Fisher, H. B., *Analyt. Chem*, 1975, **47**, 1258.
29 Fuller, C. W., *Analyt. Chim. Acta*, 1972, **62**, 261.
29a La Croix, D. E., and Wong, N. P., *Spectroscopy Letters*, 1976, **9**, 31.

30   Johnson, G. W., and Skogerboe, R. K., *Appl. Spectroscopy*, 1974, **28**, 590.
31   Hwang, J. Y., Mokeler, C. J., and Ullucci, P. A., *Analyt. Chem.*, 1972, **44**, 2018.
32   Siemer, D. D., Woodriff, R., and Watne, B., *Appl. Spectroscopy*, 1974, **28**, 582.
33   Cresser, M. S., and Mullins, C. E., *Analyt. Chim. Acta*, 1974, **68**, 377.
34   Campbell, W. C., and Ottaway, J. M., *Talanta*, 1974, **21**, 837.
35   Ellingham, H. J. T., *J. Soc. Chem. Ind.*, 1944, **63**, 125.
36   Aggett, J., and West, T. S., *Analyt. Chim. Acta*, 1971, **55**, 349.
37   Ebdon, L., Kirkbright, G. F., and West, T. S., *Talanta*, 1972, **19**, 130.
38   Aggett, J., and Sprott, A. J., *Analyt. Chim. Acta*, 1974, **72**, 49.
39   Johnson, D. J., Sharp, B. L., West, T. S., and Dagnall, R. M., *Analyt. Chem.*, 1975, **47**, 1234.
40   Posma, F. D., Balke, J., and Maessen, F. J. M. J., 18th Colloquium Spectroscopicum Internationale, Grenoble, France, September, 1975.
41   Campbell, W. C., Ph.D. Thesis, Strathclyde University, Glasgow, Scotland, 1975.
41a   Frech, W., and Cedergren, A., *Analyt. Chim. Acta*, 1976, **82**, 83.
42   L'vov, B. V., 'Atomic Absorption Spectrochemical Analysis', Adam Hilger, London, 1970.
43   Fuller, C. W., *Analyst*, 1974, **99**, 739.
44   Torsi, G., and Tessari, G., *Analyt. Chem.*, 1973, **45**, 1812.
45   Paveri-Fontana, S. L., Tessari, G., and Torsi, G., *Analyt. Chem.*, 1974, **46**, 1032.
46   Torsi, G., and Tessari, G., *Analyt. Chem.*, 1975, **47**, 839.
47   Tessari, G., and Torsi, G., *Analyt. Chem.*, 1975, **47**, 842.
48   Fuller, C. W., *Analyst*, 1975, **100**, 229.
49   Fuller, C. W., *Analyt. Chim. Acta*, 1972, **62**, 442.
49a   Fuller, C. W., *Analyst*, 1976, **101**, 798.
50   Mitchell, J. W., *Analyt. Chem.*, 1973, **45**, 492A.
51   Zief, M., and Nesher, A. G., *Environ. Sci. Technol.*, 1974, **8**, 667.
52   Belyaev, Y. I., Pchelintsev, A. M., Zvereva, N. F., and Kostin, B. I., *Zhur. analit. Khim.*, 1971, **26**, 492.
53   Belyaev, Y. I., Pchelintsev, A. M., and Zvereva, N. F., *Zhur. analit. Khim.*, 1971, **26**, 1295.
54   Langmyhr, F. J., Sturergh, J. R., Thomassen, Y., Hanssen, J. E., and Dolezal, J., *Analyt. Chim. Acta*, 1974, **71**, 35.
55   Langmyhr, F. J., Solberg, R., and Wold, L. T., *Analyt. Chim. Acta*, 1974, **69**, 267.
56   Kanda, M., Hiri, Y., and Matsumoto, I., *Bunseki Kagaku*, 1975, **24**, 299.
57   Harris, W. E., and Kratochvil, B., *Analyt. Chem.*, 1974, **46**, 313.
57a   Sommerfeld, M. R., Love, T. D., and Olsen, R. D., *Atomic Absorption Newsletter*, 1975, **14**, 31.
58   Benjamin, M. M., and Jenne, E. A., *Atomic Absorption Newsletter*, 1976, **15**, 53.
58a   Reeves, R. D., Patel, B. M., Molnar, C. J., and Winefordner, J. D., *Analyt. Chem.*, 1973, **45**, 246.
59   Maessen, F. J. M. J., Posma, F. D., and Balke, J., *Analyt. Chem.*, 1974, **46**, 1445.

60   Layman, L. R., and Hieftje, G. M., *Analyt. Chem.*, 1974, **46**, 322.
61   Pickford, C. J., and Rossi, G., *Analyst*, 1972, **97**, 647.
62   Robertson, D. E., *Analyt. Chem.*, 1968, **40**, 1067.
63   Robertson, D. E., *Analyt. Chim. Acta*, 1968, **42**, 533.
64   Struempler, A. W., *Analyt. Chem*, 1973, **45**, 2251.
65   Bernas, B., *Analyt. Chem.*, 1968, **40**, 1682.
66   Mitchell, J. W., and Nash, D. L., *Analyt. Chem.*, 1974, **46**, 326.
67   Woolley, J. F., *Analyst*, 1975, **100**, 896.
68   Gleit, C. E., and Holland, W. D., *Analyt. Chem.*, 1962, **34**, 1454.
68*a*   Cox, L. E., U.S.A.E.C., Report, LA-5791, 1974.
69   Dudas, M. J., *Atomic Absorption Newsletter*, 1974, **13**, 67.
70   Aggett, J., and West, T. S., *Analyt. Chim. Acta*, 1971, **57**, 15.
71   Clark, D., Dagnall, R. M., and West, T. S., *Analyt. Chim. Acta*, 1973, **63**, 11.
72   Gomiscek, S., Lengar, Z., Cernetic, J., and Hudnik, V., *Analyt. Chim. Acta*, 1974, **73**, 97.
73   Muzzarelli, R. A. A., and Rocchetti, R., *Analyt. Chim. Acta*, 1974, **70**, 283.
74   Newton, M. P., Chauvin, J. V., and Davis, D. G., *Analyt. Letters*, 1973, **6**, 89.
74*a*   Newton, M. P., and Davis, D. G., *Analyt. Letters*, 1975, **8**, 729.
75   Lund, W., and Larsen, B. V., *Analyt. Chim. Acta*, 1974, **70**, 299.
76   Lund, W., and Larsen, B. V., *Analyt. Chim. Acta*, 1974, **72**, 57.
76*a*   Lund, W., Larsen, B. V., and Gundersen, N., *Analyt. Chim. Acta*, 1976, **81**, 319.
77   Fairless, C., and Bard, A. J., *Analyt. Letters*, 1972, **5**, 433.
78   Fairless, C., and Bard, A. J., *Analyt. Chem.*, 1973, **45**, 2289.
79   Jensen, F. O., Dolezal, J., and Langmyhr, F. J., *Analyt. Chim. Acta*, 1974, **72** 245.
79*a*   Barnett, W. B., Vollmer, J. W., and DeNuzzo, S. M., *Atomic Absorption Newsletter*, 1976, **15**, 33.
80   Thompson, K. C., Godden, R. G., and Thomerson, D. R., *Analyt. Chim. Acta*, 1975, **74**, 289.
81   Amos, M. D., Bennett, P. A., Brodie, K. G., Lung, P. W. Y., and Matousek, J. P., *Analyt. Chem.*, 1971, **43**, 211.
82   Molnar, C. J., Reeves, R. D., Winefordner, J. D., Glenn, M. T., Ahlstrom, J. R., and Savory, J., *Appl. Spectroscopy*, 1972, **26**, 606.
83   Johnson, D. J., West, T. S., and Dagnall, R. M., *Analyt. Chim. Acta*, 1973, **66**, 171.
84   Morrow, R. W., and McElhaney, R. J., *Atomic Absorption Newsletter*, 1974, **13**, 45.
84*a*   Manning, D. C., *Atomic Absorption Newsletter*, 1976, **15**, 42.
85   Fuller, C. W., Proc. 27th BSC/BISRA Chemists Conference, Scarborough, England, May, 1974.
86   Kundu, M. K., and Prevot, A., *Analyt. Chem.*, 1974, **46**, 1591.
87   Yasuda, S., and Kakiyama, H., *Bunseki Kagaku*, 1975, **24**, 377.
88   Culver, B. R., and Surles, T., *Analyt. Chem.*, 1975, **47**, 920.
89   Adams, M. J., Kirkbright, G. F., and Rienvatana, P., *Atomic Absorption Newsletter*, 1975, **14**, 105.
90   Koirtyohann, S. R., and Pickett, E. E., *Analyt. Chem.*, 1965, **37**, 601.
90*a*   Dawson, J. B., Grassam, E., Ellis, D. J., and Keir, M. J., *Analyst*, 1976, **101**, 315.

91   Rains, T. C., Epstein, M. S., and Menis, O., *Analyt. Chem.*, 1974, **46**, 207.

92   Manning, D. C., and Fernandez, F., *Atomic Absorption Newsletter*, 1970, **9**, 65.

93   Gries, W. H., and Norval, E., *Analyt. Chim. Acta*, 1975, **75**, 289.

94   Pinta, M., and Riandey, C., 17th Colloquium Spectroscopicum Internationale, Florence, Italy, September, 1973.

94*a*   Maruta, T., Minegishi, K., and Sudoh, G., *Analyt. Chim. Acta*, 1976, **81**, 313.

95   Dolinsek, F., Stupar, J., and Spenko, M., *Analyst*, 1975, **100**, 884.

96   Yasuda, S., and Kakiyama, H., *Bunseki Kagaku*, 1974, **23**, 620.

97   Ratcliffe, D. B., Byford, C. S., and Osman, P. B., *Analyt. Chim. Acta*, 1975, **75**, 457.

98   Kamada, T., Kumamura, T., and Yamamoto, Y., *Bunseki Kagaku*, 1975, **24**, 89.

98*a*   Kunselman, G. C., and Huff, E. A., *Atomic Absorption Newsletter*, 1976, **15**, 29.

99   Frech, W., *Talanta*, 1974, **21**, 565.

100   Knudson, E. J., and Christian, G. D., *Analyt. Letters*, 1973, **6**, 1039.

101   Chapman, J. F., and Dale, L. S., 3rd Australian Symposium on Analytical Chemistry, Melbourne, Australia, May, 1975.

102   Ediger, R. D., *Atomic Absorption Newsletter*, 1975, **14**, 127.

103   Baird, R. B., and Gabrielian, S. M., *Appl. Spectroscopy*, 1974, **28**, 273.

103*a* Mesman, B. B., and Thomas, T. C., *Analyt. Letters*, 1975, **8**, 449.

104   Rozenblum. V., *Analyt. Letters*, 1975, **8**, 549.

105   Thompson, K. C., and Godden, R. G., *Analyst*, 1975, **100**, 198.

106   Renshaw, G. D., Pounds, C. A., and Pearson, E. F., *Atomic Absorption Newsletter*, 1973, **12**, 55.

106*a* Cioni, R., Mazzucotelli, A., and Ottonello, G., *Analyt. Chim. Acta*, 1976, **82**, 415.

107   Sighinolfi, G. P., *Atomic Absorption Newsletter*, 1972, **11**, 96.

108   Robbins, W. K., Runnels, J. H., and Merryfield, R., *Analyt. Chem.*, 1975, **47**, 2095.

109   Becker-Ross, H., and Falk, H., *Spectrochim. Acta*, 1975, **30B**, 253.

110   Rowe, J., and Sanders, J., Pittsburgh Conference on Analytical Chemistry and Applied Spectroscopy, Cleveland, U.S.A., March, 1973.

111   Yasuda, S., and Kakiyama, H., *Bunseki Kagaku*, 1974, **23**, 406.

112   Tominaga, M., Kimura, A., and Miyazaki, A., *Bunseki Kagaku*, 1975, **24**, 61.

113   Brodie, K. G., and Matousek, J. P., *Analyt. Chim. Acta*, 1974, **69**, 200.

114   Posma, F. D., Balke, J., Herber, R. F. M., and Stuik, E. J., *Analyt. Chem.*, 1975, **47**, 834.

115   Tessari, G., and Torsi, G., *Talanta*, 1972, **19**, 1059.

116   Maruta, T., and Takeuchi, T., *Analyt. Chim. Acta*, 1973, **66**, 5.

117   Pinta, M., and Riandey, C., *Analusis*, 1975, **3**, 86.

118   Barnard, W. M., and Fishman, M. J., *Atomic Absorption Newsletter*, 1973, **12**, 118.

119   Davidson, I. W. F., and Secrest, W. L., *Analyt. Chem.*, 1972, **44**, 1808.

120   Segar, D. A., and Gonzalez, J. G., *Analyt. Chim. Acta*, 1972, **58**, 7.

121   Hageman, L., Torma, L., and Ginther, B. E., *J. Assoc. Offic. Analyt. Chemists*, 1975, **58**, 990.

122   Riandey, C., and Pinta, M., *Analusis*, 1973, **2**, 179.

123   Yasuda, S., and Kakiyama, H., *Bunseki Kagaku*, 1974, **23** 670.
124   Everson, R. T., and Schrenk, W. G., *Appl. Spectroscopy*, 1975, **29**, 41.
124*a* Smeyers-Verbeke, J., Michotte, Y., Van den Winkel, P., and Massart, D. L., *Analyt. Chem.*, 1976, **48**, 125.
125   Vajda, F., 9th Transdanubian Analytical Conference, Szombathely, Hungary, 1973.
125*a* Everett, G. L., *Analyst*, 1976, **101**, 348.
126. Kragten, J., and Reynaert, A. P., *Talanta*, 1974, **31**, 618.
127   Maruta, T., and Takeuchi, T., *Bunseki Kagaku*, 1973, **22**, 602.
128   Shigematsu, T., Matsui, M., Fujimo, O., and Kinoshita, K., *Nippon Kagaku Kaishi*, 1973, 2123.
129   Shaw, F., and Ottaway, J. M., *Atomic Absorption Newsletter*, 1974, **13**, 77.
130   Yamamoto, Y., Kumamaru, T., Kamada, T., Tanaka, T., and Kawabe, M., *Nippon Kagaku Kaishi* 1975, 841.
131   Bratzel, M. P., and Chakrabarti, C. L., *Analyt. Chim. Acta*, 1972, **61**, 25.
132   Riner, J. C., Wright, F. C., and McBeth, C. A., *Atomic Absorption Newsletter*, 1974, **13**, 129.
132*a* Andersson, A., *Atomic Absorption Newsletter*, 1976, **15**, 71.
133   Dolinsek, F., and Stupar, J., *Analyst*, 1973, **98**, 841.
134   Ebert, J., and Jungmann, H., *Z. analyt. Chem.*, 1974, **272**, 287.
134*a* Regan, J. G. T., and Warren, J., *Analyst*, 1976, **101**, 220.
135   Shigematsu, T., Matsui, M., Fujimo, O., and Kinoshita, K., *Analyt. Chim. Acta*, 1975, **76** 329.
136   Lech, J., Siemer, D., and Woodriff, R., *Spectrochim. Acta*, 1973, **28B**, 435.
137   Lech, J., Siemer, D., and Woodriff, R., *Appl. Spectroscopy*, 1974, **28**, 78.
138   L'vov, B. V., and Khartsyzov, A. D., *Zhur. priklad. Spektroskopii*, 1969, **10**, 413.
139   Everett, G. L., West, T. S., and Williams, R. W., *Analyt. Chim. Acta*, 1974, **68**, 387.
140   Baudin, G., Bonne, R., Chaput, M., and Feve, L., 3rd International Conference on Atomic Absorption and Atomic Fluorescence Spectrometry, Paris, France, September, 1971.
141   Henn, E. L., *Analyt. Chem.*, 1975, **47**, 428.
142   Martin, T. D., Kopp, J. F., and Ediger, R. D., *Atomic Absorption Newsletter*, 1975, **14**, 109.
143   Welcher, G. G., Kriege, O. H., and Marks, J. Y., *Analyt. Chem.*, 1974, **46**, 1227.
143*a* Ihnat, M., *Analyt. Chim. Acta*, 1976, **82**, 293.
144   Ihnat, M., and Westerby, R., *Analyt. Letters*, 1974, **7**, 257.
144*a* Freedman, R. W., and Platter, D. W., 27th Pittsburgh Conference on Analytical Chemistry and Applied Spectroscopy, Cleveland, U.S.A., March 1976.
145   Fuller, C. W., *Analyt. Chim. Acta* 1976, **81**, 199.
146   Fuller, C. W., unpublished results.
147   Cioni, R., Innocenti, F., and Mazzuoli, R., *Atomic Absorption Newsletter*, 1972, **11**, 102.
148   Duewer, D. L., Kowalski, B. R., and Schatzki, T. F., *Analyt. Chem.*, 1975, **47**, 1573.
149   Eskamani, A., Vigler, M. S., Strecker, H. A., and Anthony, N. R., 4th International Conference on Atomic Spectroscopy, Toronto, Canada, October, 1973.

150   Robbins, W. K., *Analyt. Chim. Acta*, 1973, **65**, 285.
151   Kashiki, M., Yamazoe, S., Ikeda, N., and Oshima, S., *Analyt. Letters*, 1974, 7, 53.
152   Segar, D. A., *Analyt. Letters*, 1974, **7**, 89.
153   Hall, G., Bratzel, M. P., and Chakrabarti, C. L., *Talanta*, 1973, **20**, 755.
154   Araktingi, Y. E., Chakrabarti, C. L., and Maines, I. S., *Spectroscopy Letters*, 1974, **7**, 97.
155   May, L. A., and Presley, B. J., *Atomic Absorption Newsletter*, 1974, **13**, 144.
156   Omang, S. H., *Analyt. Chim. Acta*, 1971, **56**, 470.
157   Alder, J. F., and West, T. S., *Analyt. Chim. Acta*, 1972, **61**, 132.
158   Chakrabarti, C. L., and Hall, G., *Spectroscopy Letters*, 1973, **6**, 385.
159   May, L. A., and Presley, B. J., *Spectroscopy Letters*, 1975, **8**, 201.
160   Robbins, W. K., and Walker, H. H., *Analyt. Chem.*, 1975, **47**, 1269.
161   Prevot, A., and Gente, M., *Rev. Fr. Corps. Gras*, 1973, **20**, 95.
161*a*   Heinrichs, H., *Z. analyt. Chem.*, 1975, **273**, 197.
162   Robbins, W. K., *Analyt. Chem.*, 1974, **46**, 2177.
163   Oddo, N., *Riv. Combust.*, 1971, **25**, 153.
164   Everett, G. L., West, T. S., and Williams, R. W., *Analyt. Chim. Acta*, 1973, **66**, 301.
165   Brodie, K. G., and Matousek, J. P., *Analyt. Chem.*, 1971, **43**, 1557.
166   Alder, J. F., and West, T. S., *Analyt. Chim. Acta*, 1972, **58**, 331.
167   Reeves, R. D., Molnar, C. J., and Winefordner, J. D., *Analyt. Chem.*, 1972, **44**, 1913.
168   Patel, B. M., and Winefordner, J. D., *Analyt. Chim. Acta*, 1973, **64**, 135.
169   Reeves, R. D., Molnar, C. J., Glenn, M. T., Ahlstrom, J. R., and Winefordner, J. D., *Analyt. Chem.*, 1972, **44**, 2205.
170   Chuang, F. S., and Winefordner, J. D., *Appl. Spectroscopy*, 1974, **28**, 215.
171   Chuang, F. S., Patel, B. M., Reeves, R. D., Glenn, M. T., and Winefordner, J. D., *Canad. J. Spectroscopy*, 1973, **18**, 6.
172   Everett, G. L., West, T. S., and Williams, R. W., *Analyt. Chim. Acta*, 1974, **70**, 204.
173   Everett, G. L., West, T. S., and Williams, R. W., *Analyt. Chim. Acta*, 1974, **70**, 291.
174   Nikolayev, G. I., and Aleskovskii, V. B., *Zhur. analit. Khim.*, 1963, **18**, 816.
174*a*   Shaw, F., and Ottaway, J. M., *Analyt. Letters*, 1975, **8**, 911.
175   Ohta, K., and Suzuki, M., *Analyt. Chim. Acta*, 1975, **77**, 288.
176   Atsuya, I., and Suguira, N., *Bunseki Kagaku*, 1974, **23**, 1170.
177   Shaw, F., and Ottaway, J. M., *Analyst*, 1974, **99**, 184.
177*a*   Shaw, F., and Ottaway, J. M., *Analyst*, 1975, **100**, 217.
178   Frech, W., *Z. analyt. Chem.*, 1975, **275**, 353.
178*a*   Hunter, J. K., Ottaway, J. M., and Shaw, F., *Metal. Met. Form.*, 1976, 198.
179   Shrista, I. L., and West, T. S., *Bull. Soc. chim. belges*, 1975, **84**, 549.
179*a*   Maruta, T., and Takeuchi, T., *Bunseki Kagaku*, 1974, **23**, 723.
180   Frech, W., *Analyt. Chim. Acta*, 1975, **77**, 43.
180*a*   Frech, W., Lundgren, G., and Lunner, S. E., *Atomic Absorption Newsletter*, 1976, **15**, 57.
181   Sychra, V., and Kolihova, D., 2nd Czechoslovakian Flame Spectroscopy Conference, Zvikov, Czechoslovakia, 1973.
182   Nikolayev, G. I., *Zhur. analit. Khim.*, 1965, **20**, 445.
183   Ishibashi, W., Sato, M., and Hashimoto, M., *Bunseki Kagaku*, 1974, **23**, 597.
184   Medina, R., *Z. analyt. Chem.*, 1974, **271**, 346.

185   Janouskova, J., and Dudova, N., 2nd Czechoslovakian Flame Spectroscopy
      Conference, Zvikov, Czechoslovakia, 1973.
186   Knoop, P., *Analyt. Chem.*, 1974, **46**, 965.
187   Schweizer, V. B., *Atomic Absorption Newsletter*, 1975, **14**, 137.
188   Campbell, W. C., and Ottaway, J. M., *Inst. Min. Metall., Trans.*, 1974, **83**, B68.
189   Campbell, W. C., and Ottaway, J. M., *Talanta*, 1975, **22**, 729.
190   Sychra, V., Janouskova, J., Kolihova, D., Dudova, N., and Marek, S., *Chem. listy*, 1975, **69**, 623.
191   Tomljanovic, M., and Grobenski, Z., *Atomic Absorption Newsletter*, 1975, **14**, 52.
192   Brady, D. V., Montalvo, J. G., Glowacki, G., and Pisciotta, A., *Analyt. Chim. Acta* 1974, **70**, 448.
193   Bratzel, M. P., Chakrabarti, C. L., Sturgeon, R. E., McIntyre, M. W., and Agemian, H., *Analyt. Chem.*, 1972, **44**, 372.
193*a* Belyaev, Y. I., Oreshkin, V. N., and Vnukovskaya, G. L., *Zhur. analit. Khim.*, 1975, **30**, 503.
194   Heinrichs, H., and Lange, J., *Z. analyt. Chem.*, 1973, **265**, 256.
195   Riandey, C., Linhares, P. S., and Pinta, M., *Analusis*, 1975, **3**, 303.
195*a* Gong, H., and Suhr, N. H., *Analyt. Chim. Acta*, 1976, **81**, 297.
196   Cruz, R., and Van Loon, J. C., *Analyt. Chim. Acta*, 1974, **72**, 231.
197   Ohta, K., and Suzuki, M., *Talanta*, 1975, **22**, 465.
198   Beaty, R. D., *Atomic Absorption Newsletter*, 1974, **13**, 38.
198*a* Sighinolfi, G. P., *Atomic Absorption Newsletter*, 1973, **12**, 136.
199   Dudas, M. J., *Atomic Absorption Newsletter*, 1974, **13**, 109.
199*a* Osbourne, A. C., and West, T. S., *Proc. Soc. Analyt. Chem.*, 1972, **9**, 198.
200   Lundgren, G., Lundmark, L., and Johannson, G., *Analyt. Chem.*, 1974, **46**, 1028
201   Ediger, R. D., Peterson, G. E., and Kerber, J. D., *Atomic Absorption Newsletter*, 1974, **13**, 61.
202   Ediger, R. D., *Atomic Absorption Newsletter*, 1973, **12**, 151.
203   Brown, J., 4th International Conference on Atomic Spectroscopy, Toronto, Canada, October, 1973.
203*a* Gardner, D. J., Pritchard, J. A., and Sadler, M. A., *Analyst*, 1976, **101**, 278.
204   Rattonetti, A., *Analyt. Chem.*, 1974, **46**, 739.
205   Sensmeier, M. R., Wagner, W. F., and Christian, G. D., *Z. analyt. Chem.*, 1975, **277**, 19.
206   Fernandez, F. J., and Manning, D. C., *Atomic Absorption Newsletter*, 1971, **10**, 65.
207   Cernetic, J., and Gorenc, B., 5th Yugoslavian Conference on Applied Spectroscopy, Skopje, Yugoslavia, October, 1974.
208   Welz, B., *CZ Chem. Tech.*, 1972, **1**, 455.
209   Welz, B., and Wiedeking, E., *Z. analyt. Chem.*, 1973, **264**, 110.
209*a* Owens, J. W., and Gladney, E. S., *Atomic Absorption Newsletter*, 1976, **15**, 47.
210   Janouskova, J., Sulcek, Z., and Sychra, V., *Chem. listy*, 1974, **68**, 969.
210*a* Korkisch, J., Sorio, A., and Steffan, I., *Talanta*, 1976, **23**, 289.
211   Edmunds, W. M., Giddings, D. R., and Morgan-Jones, M., *Atomic Absorption Newsletter*, 1973, **12**, 45.
212   Paus, P. E., *Atomic Absorption Newsletter*, 1971, **10**, 69.
213   Briese, L. A., and Giesy, J. P., *Atomic Absorption Newsletter*, 1975, **14**, 133.

214 Maines, I. S., and Chakrabarti, C. L., 21st Canadian Spectroscopy Symposium, Ottawa, Canada, October, 1974.

215 Knott, A. R., *Atomic Absorption Newsletter*, 1975, **14**, 126.

216 Pickford, C. J., and Rossi, G., *Analyst*, 1973, **98**, 329.

217 Woodriff, R., Culver, B. R., Shrader, D., and Super, A. B., *Analyt. Chem* 1973, **45**, 230.

218 Cragin, J. H., and Herron, M. M., *Atomic Absorption Newsletter*, 1973, **12**, 37.

219 Shigematsu, T., Matsui, M., and Fujino, O., *Bunseki Kagaku*, 1973, **22**, 1162.

220 Scobie, R., *Varian Technical Topics*, 1973, (July).

221 Sato, A., Oikawa, T., and Saitoh, N., *Bunseki Kagaku*, 1975, **24**, 584.

222 Burrell, D. C., Williamson, V. M., and Meng-Lein Lee, 4th International Conference on Atomic Spectroscopy, Toronto, Canada, October, 1973.

223 Parker, C. R., and Stux, R. L., *Varian Technical Topics*, 1974, (December).

224 Shigematsu, T., Matsui, M., Fujino, O., Mitsuno, S., and Nagahiro, T., *Nippon Kagaku Kaishi*, 1975, 1328.

225 Kremlin, K., and Petersen, H., *Analyt. Chim. Acta*, 1974, **70**, 35.

226 Paus, P. E., *Z. analyt. Chem.*, 1973, **264**, 118.

227 Muzzarelli, R. A. A., and Rocchetti, R., *Analyt. Chim. Acta*, 1974, **69**, 35.

228 Muzzarelli, R. A. A., and Rocchetti, R., *Analyt. Chim. Acta*, 1973, **64**, 371.

229 Robinson, J. W., Wolcott, D. K., and Rhodes, L., *Analyt. Chim. Acta*, 1975, **78**, 285.

230 Van Loon, J. C., Lichwa, J., Ruttan, D., and Kinrade, J., *Water, Air, Soil Pollution*, 1973, **2**, 473.

231 Guillaumin, J. C., *Atomic Absorption Newsletter*, 1974, **13**, 135.

232 Baird, R. B., Pourian, S., and Gabrielian, S. M., *Analyt. Chem.*, 1972, **44**, 1887.

233 Thomas, A. D., and Smythe, L. E., *Talanta*, 1973, **20**, 469.

234 Schramel, P., *Analyt. Chim. Acta*, 1973, **67**, 69.

235 Owens, J. W., and Gladney, E. S., *Atomic Absorption Newsletter*, 1975, **14**, 76.

236 Linnmann, L., Anderson, A., Nilsson, K. O., Lind, B., Kjellstrom, T., and Friberg, L., *Arch. Environ. Health*, 1973, **27**, 45.

237 Ganje, T. J., and Page, A. L., *Atomic Absorption Newsletter*, 1974, **13**, 131.

238 Parker, C. R., Scobbie, R., and Duncan, L., *Varian Technical Topics*, 1974, (August).

239 Soerensen, N. K., *Tidsskr. Planteavl*, 1974, **78**, 156.

240 Simmons, W. J., *Analyt. Chem.*, 1975, **47**, 2015.

241 Simmons, W. J., and Loneragan, J. F., *Analyt. Chem.*, 1975, **47**, 566.

242 Fernandez, F. J., and Manning, D. C., Perkin-Elmer Atomic Absorption Application Study No. 533.

243 Manning, D. C., Perkin-Elmer Atomic Absorption Application Study No. 518.

244 Henning, S., and Jackson, T. L., *Atomic Absorption Newsletter*, 1973, **12**, 100.

245 Brady, D. V., Montalvo, J. G., Jung, J., and Curran, R. A., *Atomic Absorption Newsletter*, 1974, **13**, 118.

246 Velghe, G., Verloo, M., and Cottenie, A., *Z. Lebensm.-Untersuch.*, 1974, **156**, 77.

247 Maurer, L., *Z. Lebensm.-Untersuch*, 1974, **156** 284.

248 Kapur, J. K., and West, T. S., *Analyt. Chim. Acta*, 1974, **73**, 180.

249 Pagenkopf, G. K., Newman, D. R., and Woodriff, R., *Analyt. Chem.*, 1972, **44**, 2248.

250 Hageman, L. R., Torma, L., and Ginther, B. E., 88th Annual Meeting of the Association of Official Analytical Chemists, Washington, U.S.A., October, 1974

250*a* Slavin, S., Peterson, G. E., and Lindahl, P. C., *Atomic Absorption Newsletter*, 1975, **14**, 57.

251 Lagathu, J., and Desirant, J., *Rev. Fr. Corps Gras*, 1972, **19**, 169.

252 Roschnik, M. R., *Mitt. Geb. Lebensmittelunters. Hyg.*, 1972, **63**, 206.

253 Manning, D. C., *Amer. Lab.*, 1973, (August), 37.

254 Huffman, H. L., and Caruso, J. A., *J. Agric. Food Chem.*, 1974, **22**, 824.

255 Huffman, H. L., and Caruso, J. A., *Talanta*, 1975, **22**, 871.

256 Black, L. T., *J. Amer. Oil Chemists' Soc.*, 1975, **52**, 88.

257 Wolf, W., Mertz, W., and Masironi, R., *J. Agric. Food Chem.*, 1974, **22**, 1037.

258 Fuchs, C., Brasche, M., Paschen, K., Nordbeck, H., Quellhorst, E., and Peek, U., *Clinica Chim. Acta*, 1974, **52**, 71.

259 Welz, B., and Wiedeking, E., *Z. analyt. Chem.*, 1970, **252**, 111.

260 Aggett, J., *Analyt. Chim. Acta*, 1973, **63**, 473.

261 Ross, R. T., and Gonzalez, J. G., *Analyt. Chim. Acta*, 1974, **70**, 443.

262 Schumacher, E., and Umland, F., *Z. analyt. Chem.*, 1974, **270**, 285.

263 Ullucci, P. A., and Hwang, J. Y., 17th Colloquium Spectroscopicum Internationale, Florence, Italy, September, 1973.

264 Wright, F. C., and Riner, J. C., *Atomic Absorption Newsletter*, 1975, **14**, 103.

265 Perry, E. F., Koirtyohann, S. R., and Perry, H. M., *Clinical Chem.*, 1975, **21**, 626.

266 Cernik, A. A., and Sayers, M. H. P., *Brit. J. Ind. Med.*, 1975, **32**, 155.

266*a* Lundgren, G., *Talanta*, 1976, **23**, 309.

267 Mishima, M., *Bunseki Kagaku*, 1975, **24**, 433.

268 Machata, G., and Binder, R., *Z. Rechtsmed.*, 1973, **73**, 29.

269 Muzzarelli, R. A. A., and Rocchetti, R., *Talanta*, 1975, **22**, 683.

270 Pekarek. R. S., Hauer, E. C., Wannemacher, R. W., and Beisel, W. R., *Analyt. Biochem.*, 1974, **59**, 283.

271 Anand, V. D., and Lancaster, M. C., *Clinical Chem.*, 1974, **20**, 885.

272 Grafflage, B., Buttgereit, G., Kuebler, W., and Mertens, H. M., *Z. Klin. Chem. Klin. Biochem.*, 1974, **12**, 287.

273 Glenn. M., Savory, J., Hart, L., Glenn, T., and Winefordner, J. D., *Analyt. Chim. Acta*, 1971, **57**, 263.

274 Evenson, M. A., and Warren. B. L., *Clinical Chem.*, 1975, **21**, 619.

275 Matousek, J. P., and Stevens, B. J., *Clinical Chem.*, 1971, **17**, 363.

276 Delves, H. T., *Atomic Absorption Newsletter*, 1973, **12**, 50.

277 Delves, H. T., *Proc. Soc. Analyt. Chem.*, 1972, **9**, 268.

278 Glenn, M. T., Savory, J., Fein, S. A., Reeves, R. D., Molnar, C. J., and Winefordner, J. D., *Analyt. Chem.*, 1973, **45**, 203.

279 Olsen, E. D., Jatlow, P. I., Fernandez, F. J., and Kahn, H. L., *Clinical Chem.*, 1973, **19**, 326.

280 Yeh, Y. Y., and Zee, P., *Clinical Chem.*, 1974, **20**, 360.

281 Takeuchi, T., Yanagisawa, M., and Suzuki, M., *Talanta*, 1972, **19**, 465.

282 Ross, R. T., and Gonzalez, J. G., *Bull. Environ. Contam. Toxicol.*, 1974, **12**, 470.

283 Bek, F., Janouskova, J., and Moldan, B., *Chem. listy*, 1972, **66**, 867.

284 Bek, F., Janouskova, J., and Moldan, B., *Atomic Absorption Newsletter*, 1974, **13**, 47.

285 Slavin, S., Fernandez, F. J., and Manning, D. C., 4th International Conference on Atomic Spectroscopy, Toronto, Canada, October, 1973.

286 Zachariasen, H., Anderson, I., Kostal, C., and Barton, R., *Clinical Chem.*, 1975, **21**, 562.

287 Rosen, J. F., and Trinidad, E. E., *J. Lab. Clin. Med.*, 1972, **80**, 567.

288 Dawson, J. B., Ellis, D. J., and Fisher, G. W., 17th Colloquium Spectroscopicum Internationale, Florence, Italy, September, 1973.

289 Kubasik, M. P., and Volosin, M. T., *Clinical Chem.*, 1974, **20**, 300.

290 Ealy, J. A., Bolton, N. E., McElheny, R. J., and Morrow, R. W., *J. Amer. Ind. Hyg. Assoc.*, 1974, **35**, 566.

291 Rosen, J. F., Zarate-Salvadore, C., and Trinidad, E. E., *Pediatrics*, 1974, **84**, 45.

292 Norval, E., and Butler, L. R. P., *Analyt. Chim. Acta*, 1972, **58**, 47.

293 Hwang, J. Y., Ullucci, P. A., and Mokeler, C. J., *Analyt. Chem.*, 1973, **45**, 795.

294 Evenson, M. A., and Pendergast, D. D., *Clinical Chem.*, 1974, **20**, 163.

295 Fernandez, F. J., *Clinical Chem.*, 1975, **21**, 558.

296 Kubasik, N. P., Volosin, M. T., and Murray, M. H., *Clinical Biochem.*, 1972, **5**, 266.

297 Kubasik, N. P., Volosin, M. T., and Murray, M. H., *Clinical Chem.*, 1972, **18**, 410.

298 Hwang, J. Y., Ullucci, P. A., Smith. S; B., and Malenfant, A. L., *Analyt. Chem.*, 1971, **43**, 1319.

299 Garnys, V. P., and Smythe, L. E., *Talanta*, 1975, **22**, 881.

300 Cernik, A. A., *Brit. J. Ind. Med.*, 1974, **31**, 239.

301 Hauck, G., *Z. analyt. Chem.*, 1973, **267**, 337.

302 Kurz, D., Roach, J., and Eyring, E. J., *Analyt. Biochem.*, 1973, **53**, 586.

303 Chooi, M. K., Todd, J. K., and Boyd, N. D., *Clinical Chem.*, 1975, **21**, 632.

304 McLaughlin, E. A., Manning. D. C., and Slavin, S., Perkin-Elmer Atomic Absorption Application Study No. 540.

305 Kubasik, N. P., and Volosin, M. T., *Clinical Chem.*, 1973, **19**, 954.

306 Schaller, K. H., Essing, H. G., Valentin, H., and Schaecke, G., *Atomic Absorption Newsletter,* 1973, **12**, 147.

307 Ross, R. T., Gonzalez, J. G., and Segar, D. A., *Analyt. Chim. Acta*, 1973, **63**, 205.

308 Gross, S. B., and Parkinson, E. S., *Atomic Absorption Newsletter*, 1974, **13**, 107.

309 Christensen, S., *Atomic Absorption Newsletter*, 1972, **11**, 51.

310 Barnett, W. B., and Kahn, H. L., *Clinical Chem.*, 1972, **18**, 923.

311 Van Stenkelenburg, G. J., Van de Laar, A. J. B., and Van der Laag, J., *Clinica Chim. Acta*, 1975, **59**, 233.

312 Renshaw, G. D., Pounds, C. A., and Pearson, E. F., *Nature*, 1972, **238**, 162.

313 Pickford, C. J., and Rossi, G., *Atomic Absorption Newsletter*, 1975, **14**, 78.

313a Krishnan, S. S., Quittkat, S., and Crapper, D. R., *Canad. J. Spectroscopy*, 1976, **21**, 25.

314 Sperling, K. R., *Atomic Absorption Newsletter*, 1975, **14**, 60.

314a Blood, E. R., and Grant, G. C., *Analyt. Chem.*, 1975, **47**, 1438.

314b Ottaway, J. M., and Campbell, W. C., *Internat. J. Environ. Analyt. Chem.*, 1976, **4**, 233.

315   Evenson, M. A., and Anderson, C. T., *Clinical Chem.*, 1975, **21**, 537.

316   Lundgren, G., and Johansson, G., *Talanta*, 1974, **21**, 257.

317   Smeyers-Verbeke, J., Segebarth, G., and Massart, D. L., *Atomic Absorption Newsletter*, 1975, **14**, 153.

318   Kirkbright, G. F., and Wilson, P. J., *Atomic Absorption Newsletter*, 1974, **13**, 140.

319   Barlow, P. J., and Khera, A. K., *Atomic Absorption Newsletter*, 1975, **14**, 149.

319a   Behne, D., Bratter, P., and Walters, W., *Z. analyt. Chem.*, 1975, **277** 355.

320   Miller, R. G., and Doerger, J. U., *Atomic Absorption Newsletter*, 1975, **14**, 66.

320a   Behne, D., Bratter, P., Gessner, H., Hube, G., Mertz, W., and Rosick, U., *Z. analyt. Chem.*, 1976, **278**, 269.

321   Begnoche, B. C., and Risby, T. H., *Analyt. Chem.*, 1975, **47**, 1041.

322   Loftin, H. P., Christian, C. M., and Robinson, J. W., *Spectroscopy Letters*, 1970, **3**, 161.

323   Christian, C. M., and Robinson, J. W., *Analyt. Chim. Acta*, 1971, **56**, 466.

324   Robinson, J. W., Slevin, P. J., Hindman, G. D., and Wolcott, D. K., *Analyt. Chim. Acta*, 1972, **61**, 431.

325   Robinson, J. W., Wolcott, D. K., Slevin, P. J., and Hindman, G. D., *Analyt. Chim. Acta*, 1973, **66**, 13.

326   Robinson, J. W., and Wolcott, D. K., *Environ. Letters*, 1974, **6**, 321.

327   Robinson, J. W., and Wolcott, D. K., *Analyt. Chim. Acta*, 1973, **66**, 333.

328   Lacaux, J. P., Dinh, P. V., and Beguin, J., *Atomic Absorption Newsletter*, 1974, **13**, 49.

329   Siemer, D. D., and Woodriff, R., *Spectrochim. Acta*, 1974, **29B**, 269.

330   Siemer, D. D., Lech, J. F., and Woodriff, R., *Spectrochim Acta*, 1973, **28B**, 469.

331   Bettger, R. J., Ficklin, A. C., and Rees, T. F., *Atomic Absorption Newsletter*, 1975, **14**, 124.

332   Janssens, M., and Dams, R., *Analyt. Chim. Acta*, 1974, **70**, 25.

333   Zdrojewski, A., Quickert, N., and Dubois, L., *Internat. J. Environ. Analyt. Chem.*, 1973, **2**, 331.

334   Siemer, D., and Woodriff, R., *Analyt. Chem.*, 1974, **46**, 597.

335   Siemer, D., Lech, J. F., and Woodriff, R., *Appl. Spectroscopy*, 1974, **28**, 68.

336   Omang, S. H., *Analyt. Chim. Acta*, 1971, **55**, 439.

337   Woodriff, R., and Lech, J. F., *Analyt. Chem.*, 1972, **44**, 1323.

338   Janssens, M., and Dams, R., *Analyt. Chim. Acta*, 1973, **65**, 41.

339   Matousek, J. P., and Brodie, K. G., *Analyt. Chem.*, 1973, **45**, 1606.

339a   Noller, B. N., and Bloom, H., *Atmos. Environ.*, 1975, **9**, 505.

340   Lech, J. F., Siemer, D., and Woodriff, R., *Environ. Sci. Technol.*, 1974, **8**, 840.

341   Roques, Y., and Mathieu, J., *Analusis*, 1973, **2**, 481.

342   Hancock, S., and Slater, A., *Analyst*, 1975, **100**, 422.

343   Fuller, C. W., and Whitehead, J., *Analyt. Chim. Acta*, 1974, **68**, 407.

344   Fuller, C. W., *Atomic Absorption Newsletter*, 1975, **14**, 73.

345   Guseev, I. D., Blinova, E. S., Mayorov, I. A., and Nedler, V. V., *Zavodskaya Lab.*, 1973, **39**, 165.

346   Sherfinski, J. H., Perkin-Elmer Atomic Absorption Application Study No. 522.

347   Fuller, C. W., *Proc. Soc. Analyt. Chem.*, 1972, **9**, 279.

348   Fuller, C. W., *Atomic Absorption Newsletter*, 1973, **12**, 40.

349   Schreiber, B. E., and Frei, R. W., *Internat. J. Environ. Analyt. Chem.*, 1972, **2**, 149.

350 Jackson, K. W., West, T. S., and Balchin, L., *Analyt. Chem.*, 1973, **45**, 249.

351 Bagliano, G., Benischek, F., and Huber, I., *Atomic Absorption Newsletter*, 1975, **14**, 45.

352 Vinciguerra, G., and Thompson, B. G., *Varian Technical Topics*, 1974 (September).

353 Janouskova, J., Nehasilova, M., and Sychra, V., *Atomic Absorption Newsletter*, 1973, **12**, 161.

354 Sverdlina, O. A., Kuzovlev, I. A., Solomatin, V. S., Yurkova, V. Y., Kovykova, N. V., Nishanov, D., and Shatalina, L. G., *Zavodskaya Lab.*, 1975, **41**, 172.

355 Dittrich, K., and Zeppan, W., *Talanta*, 1975, **22**, 299.

355*a* Sherfinski, J. H., *Atomic Absorption Newsletter*, 1975, **14**, 26.

356 Vinciguerra, G., *Varian Technical Topics*, 1974 (September).

356*a* Goleb, J. A., and Midkiff, C. R., *Appl. Spectroscopy*, 1975, **29**, 44.

357 Kerber, J. D., Koch, A., and Peterson, G. E., *Atomic Absorption Newsletter*, 1973, **12**, 104.

358 Kerber, J. D., *Atomic Absorption Newsletter*, 1971, **10**, 104.

359 Dittrich, K., and Mothes, W., *Talanta*, 1975, **22**, 318.

360 Henn, E. L., *Analyt. Chim. Acta*, 1974, **73**, 273.

361 Pardhan, S. I., and Ottaway, J. M., *Proc. Analyt. Div. Chem. Soc.*, 1975, **12**, 291.

362 Surskii, G. A., and Avdeenko, M. A., *Zhur. priklad. Spektroskopii*, 1972, **17**, 564.

363 Bodrov, N. V., and Nikolaev, G. I., *Zhur. priklad. Spektroskopii*, 1974, **21**, 400.

364 Nikolaev, G. I., and Aleskovskii, V. B., *Zhur. Tekh. Fiz.*, 1964, **34**, 753.

365 Kitagawa, K., and Takeuchi, T., *Analyt. Chim. Acta*, 1973, **67**, 457.

366 Kirkbright, G. F., and Troccoli, O. E., 17th Colloquium Spectroscopicum Internationale, Florence, Italy, September, 1973.

367 L'vov, B. V., *Optika i Spektroskopiya*, 1965, **19**, 507.

368 Garton, W. R. S., and Codling, K., *Proc. Phys. Soc.*, 1960, **75**, 87.

369 Codling, K., *Proc. Phys. Soc.*, 1961, **77**, 797.

370 Brown, C. M., Naber, R. H., Tilford, S. G., and Ginter, M. L., *Appl. Optics*, 1973, **12**, 1858.

371 Vidale, G. L., General Electric, Missile and Space Vehicle Dept., T. I. S. R60SD330, 1960.

372 Nemets, A. M., and Nikolaev, G. I., *Zhur. priklad. Spektroskopii*, 1974, **21**, 405.

Index

Air particulates, analysis, 47, 105 *et seq.*
Alkali-metal halides, molecular absorption
    spectra, 56
Aluminium, conditions for determination,
    66
Antimony, conditions for determination, 66
Arsenic, conditions for determination, 66
Atomization,
    advantages and disadvantages, 18 *et seq.*
    efficiency, 9, 18 *et seq.*
    of elements, selection of conditions,
        65 *et seq.*
    of samples, selection of conditions, 51,
        84 *et seq.*
    theories, 22 *et seq.*
Atomizers,
    alignment, 50
    control units, 11, 50
    designs, 1 *et seq.*, 11 *et seq.*
      historical development, 1 *et seq.*
      parameters, experimental conditions,
        49
Automated sampling system, 21, 43

Baby food, analysis, 98
Background correction, 56 *et seq.*, 61
Barium, conditions for determination, 67
Beryllium, conditions for determination, 67
Beverages, analysis, 98
Biological fluids, 97 *et seq.*
Biological tissues, 103 *et seq.*
Bismuth, conditions for determination, 68
Blackbody radiation, from atomizers, 11,
    49, 58, 62
Blood, analysis, 100
Bone, analysis, 103
Boric oxide, analysis, 107
Bromine, conditions for determination, 68
Bullets, analysis, 109
Butter, analysis, 98

Cadmium, conditions for determination, 68
Calcium, conditions for determination, 69
Calcium carbonate, analysis, 108
Calibration methods, 42, 59 *et seq.*
Carbide formation, 26, 54, 55, 63

Catalysts, analysis, 109
Cephalin, analysis, 103
Cerebrospinal fluid, analysis, 102
Cheese, analysis, 98
Chromium,
    analysis, 87
    conditions for determination, 69
Clean air conditions, use of, 41
Coal, analysis, 109
Cobalt, conditions for determination, 70
Coffee powder, analysis, 98
Copper,
    analysis, 87
    conditions for determination, 70
Cup atomizer,
    advantages for analysis of solids, 42
    for analysis of atmospheric
        particulates, 47, 107

Designs,
    commercial atomizers, 11 *et seq.*
    graphite cup, 2, 14
    graphite furnace, 1 *et seq.*, 11
    graphite rod, 7, 8, 14, 15
    tantalum filament, 7, 16
Detection limits,
    attained for elements, 65 *et seq.*
    comparison of electrothermal
        atomizers, 21
    comparison with flame atomization,
        18
Drugs, analysis, 97

Eggs, analysis, 103
Electrodeposition, as a separation
    technique, 48
Ellingham diagrams, 23
Emulsions, as a method of sample
    preparation, 46
Erbium, conditions for determination,
    71
Europium, conditions for determination,
    71

Faeces, analysis, 103
Finger nails, analysis, 103

Fish, analysis, 98
Food, analysis, 97
Fume extraction from atomizers, 13

Gallium, conditions for determination, 71
Gallium arsenide, analysis, 109
Germanium, conditions for determination, 71
Glass, analysis, 108
Gold,
    analysis, 87
    conditions for determination, 72
Grain, analysis, 98
Graphite tubes, grooved, advantages, 12
Gunshot residues, analysis, 110

Hair, analysis, 103
Hydrocarbon gases, use of as a sheathing gas
    for graphite atomizers, 54
Hydrocarbon diffusion flame, use of as a
    sheathing gas, 53

Ilmenite, analysis, 90
Indium, conditions for determination, 72
Inert gases, use of as sheathing gas, 52
Interferences,
    chemical, 62
    effect of hydrogen diffusion flame, 53
    physical, 61
    theoretical studies, 26, 36
Iodine, conditions for determination, 72
Ion implantation, technique for preparing
    solid standards, 60
Iridium, conditions for determination, 73
Iron,
    analysis, 87
    conditions for determination, 73
Iron ore, analysis, 90

Kinetic theories of atomization, 27 et seq.
    analytical applications, 31 et seq.

Lead, conditions for determination, 73
Light scattering, 55, 56
Light sources, 49, 62
Lining, metal, for graphite atomizers, 3, 13
Lithium, conditions for determination, 74

Magnesium, condition for determination, 74
Manganese, conditions for determination, 74
Matrix, condensation during atomization,
    13, 14, 63
Matrix control, theoretical, 32
Meat, analysis, 98
Memory effects, 52, 62

Mercury,
    analysis, 88
    conditions for determination, 75
Metals, analysis, 86 et seq.
Micropipettes, 43
Milk, analysis, 98
Minerals, analysis, 88 et seq.
Molecular absorption, 56
Molybdenum,
    analysis, 88
    conditions for determination, 75

Nickel,
    analysis, 88
    conditions for determination, 75
Niobium, analysis, 88
Nitride formation, 64

Oil products, 84 et seq., 99
Optical apertures, 11, 14, 16, 49
Organic solvents, effect on calibration,
    12, 48, 59
Osmium, conditions for determination,
    76
Oxides,
    evaporation, 22
    reduction, 23
    thermal dissociation, 22
Oxygen, used as a sheathing gas, 55

Palladium, conditions for determination,
    76
Paper, analysis, 110
Phosphorus, conditions for
    determination, 76
Plants, analysis, 96
Plasma, analysis, 100
Plastics, analysis, 110
Platinum, conditions for determination,
    77
Plutonium, conditions for determination,
    77
Potassium, conditions for determination,
    77
Pressure vessels, use, 46
Pyrolysis,
    of elements, selection of conditions,
        65 et seq.
    of samples, selection of conditions,
        51
    loss of elements during, 62
Pyrolytic graphite coatings, 4, 17, 54
    apparatus for forming, 17
    continuous formation, 54
    effect on sensitivity, 55

Rare-earth oxides, analysis, 108
Reaction mechanisms, theoretical, 27, 38
Reagents, chemical, purity, 41, 46
Refractory oxides, analysis, 107
Rhenium, conditions for determination, 78
Rhodium, conditions for determination, 78
Rocks, analysis, 88 *et seq.*
Rubidium, conditions for determination, 78
Ruthenium, conditions for determination,
    78

Sample contamination, 41, 44
Samples,
    gaseous, analysis, 47
    handling, 41 *et seq.*
    in situ treatment, 19
    liquid, analysis, 43 *et seq.*
    radioactive, analysis, 47
    selection of conditions for analysis, 49
    solid, analysis, 42
Sediments, sea, analysis, 90
Selenium, conditions for determination, 79
Sensitivity,
    attained for elements, 65 *et seq.*
    comparison with flame atomization, 18
    effect of pyrolytic graphite coatings, 54
Separation procedures, use of, 47
Serum, analysis, 100
Sheathing gas, selection, 52 *et seq.*
Shellfish, analysis, 98
Signal measurement, peak height/
    integration, 32, 60
Silica, analysis, 108
Silicon,
    analysis, 110
    conditions for determination, 79
Silicon nitride, analysis, 110
Silver,
    analysis, 88
    conditions for determination, 80
Snow, analysis, 94
Soap, analysis, 110
Sodium, conditions for determination, 80
Sodium carbonate, analysis, 108
Soils, analysis, 88 *et seq.*

Stopped gas-flow, use of, 12, 32, 52
Strontium, conditions for determination,
    80
Sugar, analysis, 99
Sulphur, conditions for determination, 81
Support electrodes, regeneration of
    contact surfaces on, 14
Suspensions, as a method of sample
    preparation, 46
Sweat, analysis, 102
Synthetic protein, analysis, 99

Tantalum, analysis, 88
Tea powder, analysis, 99
Teeth, analysis, 105
Tellurium, conditions for determination,
    81
Thallium, conditions for determination,
    81
Theoretical parameters, measurement,
    110
Thermodynamic theories of atomization,
    22 *et seq.*
    analytical applications, 26
Tin,
    analysis, 88
    conditions for determination, 82
Titanium, conditions for determination,
    82
Tungsten, analysis, 88

Uranium oxide, analysis, 109
Urea, as a soild standard, 60
Urine, analysis, 102

Vanadium, conditions for determination,
    82

Water, analysis, 92 *et seq.*

Yeast, analysis, 105

Zinc, conditions for determination,
    83